Edge Intelligence

Deep learning-enabled edge computing

Online at: https://doi.org/10.1088/978-0-7503-5593-3

IOP Series in Next Generation Computing

Series editors

Prateek Agrawal
University of Klagenfurt, Austria and Lovely Professional University, India

Anand Sharma
Mody University of Science and Technology, India

Vishu Madaan
Lovely Professional University, India

About the series

The motivation for this series is to develop a trusted library on advanced computational methods, technologies, and their applications.

This series focuses on the latest developments in next generation computing, and in particular on the synergy between computer science and other disciplines. Books in the series will explore new developments in various disciplines that are relevant for computational perspective including foundations, systems, innovative applications, and other research contributions related to the overall design of computational tools, models, and algorithms that are relevant for the respective domain. It encompasses research and development in artificial intelligence, machine learning, block chain technology, quantum cryptography, quantum computing, nanoscience, bioscience-based sensors, IoT applications, nature inspired algorithms, computer vision, bioinformatics, etc. and their applications in the areas of science, engineering, business, and the social sciences. It covers a broad spectrum of applications in the community, including those in industry, government, and academia.

The aim of the series is to provide an opportunity for prospective researchers and experts to publish works based on next generation computing and its diverse applications. It also provides a data-sharing platform that will bring together international researchers, professionals, and academics. This series brings together thought leaders, researchers, industry practitioners, and potential users of different disciplines to develop new trends and opportunities, exchange ideas and practices related to advanced computational methods, and promote interdisciplinary knowledge.

A full list of titles published in this series can be found here: https://iopscience.iop.org/bookListInfo/iop-series-in-next-generation-computing.

Edge Intelligence

Deep learning-enabled edge computing

Shajulin Benedict
Indian Institute of Information Technology Kottayam, Kerala, India

IOP Publishing, Bristol, UK

ISBN 978-0-7503-5593-3 (ebook)
ISBN 978-0-7503-5591-9 (print)
ISBN 978-0-7503-5594-0 (myPrint)
ISBN 978-0-7503-5592-6 (mobi)

DOI 10.1088/978-0-7503-5593-3

Version: 20240601

IOP ebooks

British Library Cataloguing-in-Publication Data: A catalogue record for this book is available from the British Library.

Published by IOP Publishing, wholly owned by The Institute of Physics, London

IOP Publishing, No.2 The Distillery, Glassfields, Avon Street, Bristol, BS2 0GR, UK

US Office: IOP Publishing, Inc., 190 North Independence Mall West, Suite 601, Philadelphia, PA 19106, USA

To my father C Benedict, mother J Mary Alphonsal, wife Rejitha R S,
son Shaun Benedict, and daughter Sharlina R S.

Contents

Part IV Applications

Preface

'The day the power of love overrules the love of power, the world will know peace.'—Mahatma Gandhi

'At the heart of silence is prayer. At the heart of prayer is faith. At the heart of faith is life. At the heart of life is service.'—Mother Teresa

Applying intelligence to sensor data using clouds has evolved with multiple productive innovations in the recent past. Edge-enabled computing has captivated the minds of several researchers, academicians, and practitioners. A few crucial questions that arise as a paradigm shift among practitioners are listed as follows:

- How effectively can we protect the privacy of data while incorporating intelligence into applications?
- Can we engage multiple machines or services to develop a robust learning model, especially considering the resource-constrained edge-level nodes?
- How can edge-level intelligence solve the grassroots problems of societies?

In fact, solving known challenges and developing learning models using the traditional approach is not a viable solution in the modern world of demanding requirements. It is important to understand the basic principles of edge-enabled intelligence and security features.

This presents an inspirational note on how to apply deep learning intelligence using tiny edge-level computing nodes. It enables researchers and practitioners to apply the tools and principles of edge-level intelligence in their specific problem domains, such as healthcare, transportation, industry, and so forth. It draws upon the attraction of readers to practice with innovative ideas that could be developed in the near future.

I welcome the readers of this book to learn and rapidly evolve with some research to product-level developments that benefit society at large. I wish that you spread your innovation wings across the globe based on the insights gained from the book.

I dedicate this book to my loving parents (Mr C Benedict and Mrs J Mary Alphonsal), wife (Rejitha R S), and children (S Shaun Benedict and Sharlina R S). I acknowledge my mentor, Professor Michael Gerndt of the Technical University Munich, Germany, since 2019. In addition, I extend my gratitude to AIC/IIIT-Kottayam, the editors and reviewers of this publishing house.

—Dr Shajulin Benedict
(http://www.sbenedictglobal.com)

Acknowledgements

I acknowledge the support provided by Professor Michael Gerndt of TUM, IIIT-Kottayam well-wishers, Government of India, and my fellow PhD scholars for supporting this work.

Author biography

Shajulin Benedict

Shajulin Benedict graduated in 2001 from Manonmaniam Sunderanar University, India, with Distinction. In 2004, he received his ME Degree in Digital Communication and Computer Networking from A.K.C.E, Anna University, Chennai. He is the University second rank holder for his masters. He did his PhD degree in the area of grid scheduling at Anna University, Chennai (Supervisor: Dr V Vasudevan, Director, Software Technologies Group of TIFAC Core in Network Engineering). He was affiliated with the same group and published further papers in international journals.

After gaining his PhD, he joined a research team in Germany to pursue PostDoctorate research under the guidance of Professor Gerndt. He served as Professor at the SXCCE Research Centre of Anna University-Chennai. Later, he visited TUM Germany to teach cloud computing and Internet of Things as a Guest Professor at TUM-Germany.

Currently, he works at the Indian Institute of Information Technology Kottayam, Kerala, India, an institute of national importance in India, and as a Guest Professor at TUM-Germany. Additionally, he serves as Director/PI/Representative Officer of AIC-IIITKottayam (Sec.8 Company) for nourishing young entrepreneurs in India. His research interests include deep learning, HPC/cloud/grid scheduling, performance analysis of parallel applications (including exa-scale), and IoT cloud.

Part I

Introduction

Chapter 1

Edge intelligence

Edge computing is a computing paradigm in which a few computational resources are arranged nearer to data sources. It promotes robust computing architectures to perform in novel ways in collaborative learning processes. Edge computing has motivated several researchers and practitioners from various domains such as healthcare, agriculture, industry, education, finance, etc, to apply edge computing essentials, i.e., pushing services from cloud-level networks to edge-level end networks. This chapter explains the basics of edge computing and provides an introduction to edge-level intelligence applied using deep learning algorithms.

1.1 Edge computing

A modern Internet of Things (IoT)-enabled application consists of compute/storage units dispersed across various geographical locations, including the resources at the peripheral end, i.e., nearer to the sensor nodes. These compute/storage resource units are often tiny and computationally less powerful than distanced cloud nodes.

Computations that occur on peripheral devices that are nearer to the data sources are termed edge computing. It is an extended taxonomy of traditional distributed computing paradigms. Edge computing enables the creation of modest decentralized networks without including clouds or centralized processing nodes.

In fact, edge computing has created a formidable opportunity for researchers in various areas. For instance, enterprises apply edge computing [10] to increase business capabilities while attracting and processing their client's data nearer to them; robotic devices utilize edge computing [15] to quickly process autonomous decisions similar to self-driving cars, etc. Most importantly, applying intelligence to data-intensive IoT-enabled applications has challenged tens of thousands of researchers with numerous research problems.

doi:10.1088/978-0-7503-5593-3ch1

1.1.1 Objectives of edge computing

The most important reasons for incorporating edge computing in IoT-enabled applications are listed as follows:

1. Hardware cost: In recent years, the cost of commodity hardware units has drastically dropped. Developers have notions of creating software or intelligent solutions based on these devices at cheaper prices. For instance, the cost of a Raspberry Pi is less than INR 4000. Obviously, a cluster of Pis could be created to increase the performance while applying intelligence.
2. Privacy and security: Designers have enhanced the privacy of edge-enabled applications. For instance, human data such as their emotions are not submitted to cloud resources in order to study the behavior of human characteristics if edge computing is utilized; similarly, data from mobile phones are secured in an edge computing solution as learning models are transferred to centralized learning models in a distributed learning platform.
3. Network bandwidth: Edge computing can reduce the network bandwidth of sensor-based applications [9]. This is because instead of sending the data to clouds or data centers, they are processed closer to the data sources. The majority of network users have benefited greatly from it as networking systems' end-throughput in edge computing systems has increased.
4. Avoid vendor lock-in: Edge computing can mitigate serious vendor lock-in threats of entire cloud-based solutions. Although modern execution environments such as serverless computing have emerged recently, the vendor lock-in issue has remained because of in-house architectures. Notably, edge-based implementations pave the way for orchestrating computations on peripheral nodes belonging to multiple vendors, including their resources.

   The vendor lock-in problem has created severe detrimental effects on economically growing startups. For instance, the problem could create increased costs, disrupted businesses/services, and unpredicted future costs to the startup founders.
5. Energy/performance efficiency: Edge computing can offer a higher performance/energy efficiency to IoT-enabled applications. This feature is much higher in cases of processing concurrent tasks or programs. In general, edge computing nodes cannot execute large-scale computational programs. Handling such tasks requires an elegant approach to dealing with them.
6. Big data processing: Most recent applications from the computer vision research domain have a potential surge in input data to computational nodes. Grabbing all these data on centralized servers is not considered an optimal solution. Edge computing, due to its decentralized networks, delivers a better solution for processing data. For instance, data servers could be decentralized nearer to the edges such that only limited data are necessary in cloud-based data centers. Figure 1.1 illustrates the objectives of incorporating edge computing in AI-specific applications.

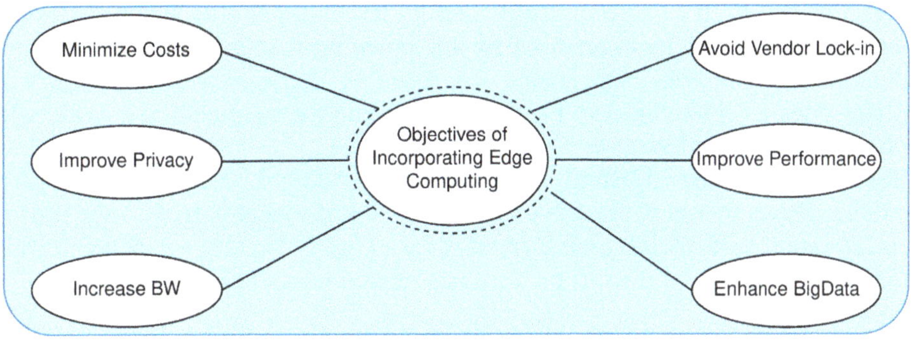

Figure 1.1. Objectives of edge computing.

1.2 History of edge computing

Edge computing evolved based on the previous concept of clouds, named content delivery networks (CDNs). In fact, the evolution of CDNs dates back to the 1990s, when a portion of bandwidth-heavy video and data contents were processed in localized regions.

The core concept of processing data nearer to data sources was reintroduced as cloudlets in 2009. In the cloudlet approach, the emphasis was not only on content but also on a minimal version of computational units nearer to data sources. The cloudlets were termed a mini-cloud datacenter, with more emphasis on enriching mobile applications.

In later years, around 2012, the concept of involving clustered servers was introduced in the name of fog computing clusters for sensor-enabled applications. These clusters are often hosted in city centers or factory locations. The fog computing units are powerful devices that led to expensive installations.

Overriding the challenges of installing a pure fog cluster, the concept of edge computing evolved, which assisted offline and online modes of communication between end devices and compute nodes.

1.3 Edge intelligence

One of the core objectives of utilizing edge computing is to learn, infer, and act on edges. In this way, applications can gain near-real-time insights on data and perform decisive actions based on edge-level processed data. Edge intelligence is deployed in two broad ways: (i) machine learning-based intelligence; and (ii) deep learning-based intelligence.

1.3.1 Machine learning-based edge intelligence

In this category, machine learning algorithms are deployed on edge nodes or devices to capture, learn, assess, and decide on key activities. These tasks need to be agile without any compromise on expected learning qualities.

Edge devices need to automatically identify learning patterns, detect anomalies, and modify data behavior in real time by implementing machine learning algorithms and models at the edge. Obviously, the speed of decisions will be faster while applying edge-level intelligence. For instance, predictive maintenance is faster while applying edge-enabled systems [2].

Traditional machine learning models such as random forests, support vector machines, Bayes theorem, etc, have limitations in processing data, i.e., the learning accuracy reduces with increasing input data. This is because machine learning algorithms are prone to overfitting and bias-related issues in many use cases.

1.3.2 Deep learning-based edge intelligence

Deep learning, in general, is a sub-field of machine learning that mimics the learning processes of humans. Our human brains learn different data based on several histories of information. This involves computationally powerful computers or computing devices to learn a large volume of data that arises from data-intensive applications, such as IoT-enabled applications.

In recent years, the majority of real-world applications have included deep learning algorithms on edge nodes, considering the performance efficiency and learning accuracy with respect to the input regional data [6, 11, 14]. The most widely applied deep learning algorithms are convolutional neural networks (CNNs), image segmentation algorithms, generative adversarial networks (GANs), reinforcement learning (RL), and transformers. The application of deep learning algorithms on edge-level nodes increases privacy and avoids latency while considering a large chunk of data in an automated fashion.

1.3.3 Hype on edge intelligence

Edge computing and the application of deep learning algorithms to edge nodes have overwhelmed researchers in the recent past. Tens of thousands of researchers have contributed to the growth of these topics. In the recent past, there has been a steep rise in the number of publications. For instance, for the keyword 'edge intelligence', the number of publications reached over 30 000 in 2023 when compared to 2637 in 2009 (see figure 1.2).

In addition, there are over 165 000 granted patents for the keyword 'edge intelligence' as of February 2024 by companies such as IBM, Samsung, Dell, Intel, AT&T, etc, apart from academic patents [7].

1.4 Advantages: edge intelligence

The hype surrounding edge intelligence is so great that various deep learning algorithms have been integrated into edge-based computing environments. The potential growth factors of edge intelligence are multi-faceted, as discussed below.

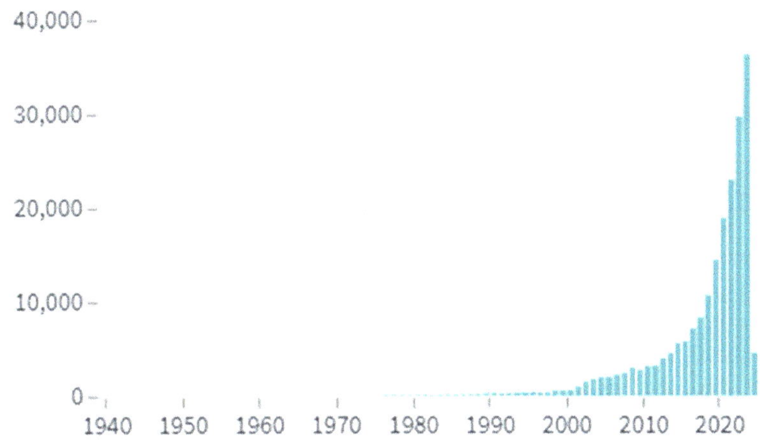

Figure 1.2. Publications for the keyword 'Edge Intelligence'. Reprinted with permission from https://www.lens.org/.

1.4.1 Low latency

Applying edge intelligence to applications delivers faster responses to provide realistic smart applications. In general, edge intelligence is meant to offer decision-making processes closer to the sources of data. For instance, a user with a mobile phone can check health parameters such as heart rate, temperature, etc, based on the features of the application. The user can view real-time data and a history of processed data in a visual form. All these applications are often executed using edge-level compute nodes.

As can be seen, the data does not need to transit to cloud services hosted on third-party servers that are decentralized, involving a combination of edge, fog, and cloud servers.

1.4.2 Enhanced security

Edge computing, in general, enables security for data and hardware components. For instance, data are processed nearer to sensors, i.e., these data are doomed to reside within the network administration where devices are located. This feature strengthens the security feature while applying deep learning algorithms or a few distributed learning techniques, such as federated learning, in specialized frameworks.

1.4.3 Bandwidth efficiency

Providing edge-level intelligence reduces communications between cloud devices and the other hierarchical nodes. In traditional learning approaches, sensor data is transported and stored in cloud storage units. The affinity of data to the storage units is not controlled by cloud users. Instead, they are configured and deployed using cloud providers or infrastructure providers.

Due to the provision of edge-based control units, there is a higher chance of positioning data on irrelevant distributed cloud storage units, depending on the availability of memory spaces. This action can lead to increased bandwidth in networks that carry sensor data for processing or storing it on cloud units.

1.5 Challenges: edge intelligence

Despite several enthusiastic approaches to applying edge-level intelligence, the concept of edge intelligence has a few drawbacks or challenges. The most commonly found challenges are listed below:

- Heterogeneity: Edge-level devices are heterogeneous in nature. It varies from architectural changes to programming-level variations. For instance, edge computing devices consist of moderately powerful computational servers or battery-operated microcontroller devices. The processing capabilities and storage capacities significantly vary among devices. Thus, the underlying computational variations are considered a crucial challenge to developing a holistic edge-intelligent system.
- Configurational settings: Sensor-enabled edge gateways or devices can be configured using various parameters. For instance, Espressif's ESP processor-based board is configured by switching ON or OFF Bluetooth or WiFi components. Notably, the performance and energy consumption values of devices vary. Hence, there is an urgent need to capture the appropriate configurational settings for better efficiency in handling edge intelligent-based tasks.
- Big Data processing: Storing a large volume of emerging data from IoT-enabled sensor sources or elegantly processing them considering the limitations of the available memory capacities of edge nodes is considered a challenge while dealing with edge intelligence.
- Diverse input streams: Sensor data processed at edge nodes can have multi-modal input data to process certain applications. For instance, edge devices that need to process voice assistant applications have to handle audio files; a few devices must process video streams so that images are efficiently handled.
- Machine failures: There is a higher probability of device failures in small, commodity hardware components. These devices run out of memory space or have component failures, such as WiFi or BLE non-availability. Obviously, it becomes an arduous task for associated applications or schedulers to organize machines in an efficient manner.
- Data hiding/tampering: Having all data sources and processing elements on edge devices makes them susceptible to data tampering or data hiding problems. Data tampering is a process in which end devices or human operators attempt to modify the contents of data sources. In doing so, appropriate data is not processed on the cloud servers or application services. Hiding data is a process in which some data is not exposed to hierarchical nodes or processing elements. Therefore, the intended output varies from the real situation.

1.6 Applications

Applying deep learning or allied learning algorithms to edge devices has been practiced in various domains. The emergence of edge-enabled applications has significantly increased in recent years. Some popular edge-enabled applications that have potential applications for deep learning algorithms are classified into several categories, as detailed below.

1.6.1 Retail sector

Selling goods to customers located in different parts of a country or a state is a tiresome task. A large volume of data emerges from millions of people that needs to be processed nearer to edge nodes. For instance, identifying the customer base, studying the requirements and needs of customers, automatic order placements, etc, have increased the application of intelligence, especially using deep learning algorithms, in edge nodes.

1.6.2 Agriculture sector

There have been efforts in the past to make decisions on remote agricultural farms using edge nodes without the assistance of cloud servers. Notably, tracking farms [5] and observing the health status of crops have attracted the involvement of edge nodes instead of clouds. There are tens of thousands of use cases that utilize IoT-enabled sensor nodes for managing agricultural solutions. However, the interest in applying edge-enabled intelligence to existing solutions in agricultural sectors has been of great interest to a number of researchers in recent years.

1.6.3 Healthcare sector

Applying edge intelligence to healthcare applications is an appropriate approach owing to the requirement of providing privacy to patients while offering intelligence [13]. In the healthcare sector, many applications that control the attached actuators or that require immediate local responses are considered potential use cases. For instance, fall detection solutions in the healthcare sector, where elderly people have to be monitored to prevent unnecessary bone fractures due to falling down, consist of actuators such as air balloons fitted to textiles or wearables. These balloons must be activated immediately when there is a change in the angle beyond a specific threshold during the movement of the body. Several authors have developed an age estimation approach to correlate the old-age fall detection systems [8].

1.6.4 Education sector

In the education sector, teachers have to explain concepts from multiple perspectives. In doing so, interpreting the concepts will become easier. In this context, IoT-enabled virtual reality-based solutions have been developed. Students can virtually feel and touch human organs, learn to play tennis, react to robotic questions, etc, based on IoT-enabled devices. To achieve better results and response times, edge-based intelligence is applied to these solutions. Researchers have demonstrated the use of edge computing to improve data transfers to clouds [12]. In the future, there is the possibility to develop educational robots that create model teachers or professors by collecting necessary novel information on subjects.

1.6.5 Industrial sector

Edge intelligence is quite commonly applied in the industrial sector. Industries have numerous machines that need to collect minute information from locations inaccessible to humans. Also, machines must communicate their status to other machines for automatic processes. In some cases, machines need to be trained by applying an intelligent ecosystem of machines, e.g., digital twins. Obviously, edge intelligence plays a vital role in this sector, as there is a need to process intelligence and offer timely decisions. Additionally, researchers and practitioners created a subdivision called industrial edge computing to innovate with ideas that accelerate industrial processes and create revenue in multiple ways. For instance, the authors of [4] developed a mechanism to place containers that could efficiently minimize the costs involved in resource utilization.

1.6.6 Finance sector

The major dimension of implementation in the finance sector is focused on offering security to financial communications. Providing real-time financial insights, procuring international trading components online, developing personalized financial dealings, etc, have become lucrative models with the assistance of edge-based solutions. In the financial sector, security-oriented solutions are crucial factors to enable a game-change in the existing financial market. Devices with near-field communications that automatically offer payments based on credit cards must be secured in all transactions. The future solutions must have direct communications and intelligence based on edge devices.

1.6.7 Smart city sector

Intelligent solutions are incorporated into smart city solutions to manage freebies, provide smart parking lots [1], increase the revenue of citizens, offer smart lighting conditions, promote happiness among citizens, etc. In fact, assessing the emotions of citizens using edge devices is a better solution than transferring images of citizens to third-party cloud environments. Some applications in the smart city sector are well established in the fields of bare IoT systems. However, significant projects are yet to start, considering the application of edge intelligence.

1.6.8 Transportation sector

In all transportation mediums, such as rail, air, road, and water, edge intelligence plays a major role. For instance, signaling on railway tracks, controlling travelers, guiding passengers, flight planning, logistics planning, etc. In fact, in the future, most robotic machines or devices will have the capability to automatically make decisions based on historical information. Researchers, therefore, are interested in exploring the possibilities to improve the efficiency of securely handling these tasks.

It is also expected that smart cooperative transportation systems will evolve in the future. Smart mobility solutions considering the air quality of streets or nature-curative pollution levels will become a practical norm in the coming years [3].

1.6.9 Gaming sector

The number of video gamers increases day by day as social involvement among people has decreased in the modern world. The gaming industry has started to improve upon the requirements of gamers. In general, collaborative gaming, which engages multiple players from decentralized geographical locations, requires game streaming, virtual reality, and allied technologies such as sensor technologies to effectively deal with the play rules of gamers.

In fact, applying variants of deep learning to the edge devices of players can increase gaming experiences in addition to routine processes. Gamers will experience near-real-time play with the involvement of audio, video, and textual descriptions.

1.6.10 Personal sector

Pervasive computing emerged in the past to enrich personal activities using gadgets such as cameras, smartphones, voice assistants, etc. The majority of users have therefore started to utilize a larger number of devices to keep them productive during their work. For instance, a book writer chooses to involve devices that will remind them about recent topics, semantics of themes, word suggestions, grammar correction engines, etc, nearer to the work office; similarly, a homemaker will choose devices to automatically inform them about the availability of purchasing items from groceries or stores. It is mandatory to assess the requirements of personal characteristics before opting for decisions in this domain. Obviously, devices that will automatically pinpoint the best solutions occur through the incorporation of edge-based intelligent solutions.

Figure 1.3 provides a recap of the massive application-related sectors in which deep learning-assisted edge computing could be applied in a pictorial form.

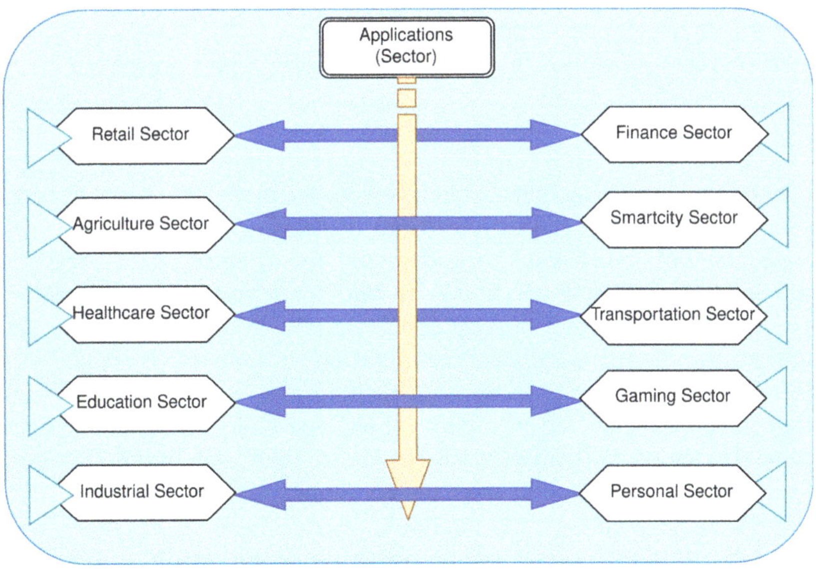

Figure 1.3. Sectors of edge-AI applications.

1.7 The need for this book

The motivation behind writing this book is to reveal the importance of applying deep learning on edge nodes. Many researchers have come to the conclusion that the cloud is not sufficient for performing deep learning applications. In this context, we must know the inner details of edge computing-based techniques for IoT-enabled applications.

This book will broaden the knowledge and thought processes of researchers, students, industrialists, practitioners, and faculties in developing edge-enabled applications. It places emphasis on IoT-enabled applications that involve deep learning algorithms.

The book is crucial for next-generation IoT-enabled application developers and practitioners for the following reasons:

1. It delivers a number of edge-enabled techniques based on deep learning approaches that focus on image classification, image enhancement, word analysis, human–machine emotional interfaces, etc.
2. It explores the techniques involved in edge computing, such as model compression, transfer learning, federated learning, energy efficiency aspects of distributed intelligence, etc.
3. It provides platforms and techniques that could be incorporated by developers or researchers. The platforms that are discussed here could be extended by readers by incorporating their inherent ideas (specified in the book).
4. It explains the importance of different edge devices, including edge-level accelerators, to increase the performance of applications.
5. It provides edge-based deep learning techniques that offer accurate decisions on inferences.
6. Additionally, a few applications are explained which reveal sample use cases for future developers. Applications that focus on elephant emotion analysis and business portfolios that emphasize societal uplift are given.

1.8 Potential readers

I wrote the book considering diverse readers from the global market. These potential readers may have specific perspectives about the book.

- Students: They may want to understand the application of edge nodes for developing intelligent solutions. In fact, meta-verse applications or social good applications require edge computations. Students will understand the importance of such edge-level computations in innovative applications.
- Researchers: They may have to read this book to learn existing techniques that have been handled by many other researchers and industrialists in the past. The book will provide deeper knowledge and insights to undertake future types of research in the edge computing domain.
- Practitioners: While reading the book, practitioners will get more innovative ideas as it pinpoints several examples touching upon newer ideas and solutions for developing edge-intelligent products.

- Government/smart city officials: The techniques provided here could provide guidance for best practice for smart city officials. For instance, they will get insights on maintaining privacy while learning the behavior of their citizens at the edge nodes; they will understand how to utilize specific edge accelerators for their governmental projects.

1.9 Organization of the book

The book is organised into four major parts. Part A contains chapters 1–3. In this part, the most important basics that lead to understanding edge intelligence, such as IoT, architectures, traditional ML/DL, etc, are discussed. Part B delves into the core concepts of developing edge-level intelligence using edge-assisted nodes, including accelerators such as Jetson Nano. Chapters 4–8 deal with the learning aspects and the corresponding performance or security features that a reader of this book needs to understand. I have attempted to discuss as many techniques as possible covering edge-level intelligence so that readers can obtain deeper insights into the existing techniques. Part C, chapters 9 and 10, describe the tools and platforms that provide edge-level intelligence using deep learning or similar learning algorithms. Providing such existing tools is to enable readers to develop newer techniques by enhancing or incorporating the tools for their problem domains. The remaining chapters appear in Part D where emphasis is placed on the applications that are widely utilized in various domains, with suitable examples. This part provides a few suggestions that students, researchers, or project enthusiasts could undertake in their careers.

In order to engage readers throughout the chapters, I have written the book in a specific organizational style. The synopsis of the chapters and their core ideas are summarized below:

1. Edge computing: Introduction: This chapter provides a brief introduction to the basics of edge computing. It highlights the importance of edge nodes and provides a synopsis of the book.
2. Edge architectures: In general, next-generation computing applications require sophisticated architectures, mostly heterogeneous compute nodes. In the second chapter, I explain the basics of computing and IoT technologies that remain the basics of understanding edge-level intelligence. Next, I introduce the edge-level architectures that have the capability to incorporate deep learning algorithms for providing intelligence.
3. Edge-level OS and programming models: Edge devices are often battery-operated or energy-conscious. The most commonly available OS, such as Ubuntu or Windows, cannot be applied to such devices. This chapter explains the edge-oriented OS (namely, FreeRTOS) and its features for developing parallel tasks in edge AI applications. In addition, it describes the firmware development procedures and/or booting processes of micro-controllers. These concepts will be beneficial for readers who dive into system-level understanding while developing edge-AI applications.
4. Learning techniques: Typically, learning at the edge happens in several ways. For instance, transfer learning, federated learning, gossip learning,

etc, are some of the popular techniques in this field of interest. In this chapter, I explain the most notable techniques that are utilized for learning or modeling using edge nodes.

5. Inferencing techniques in edge nodes: In this chapter, I explain the inferencing techniques of AI applications. Inferencing, in general, involves pre-processing techniques, layering of neural networks, and algorithms. The chapter explains a number of components, such as cameras and tools, for inferencing data. Additionally, it lists the procedure to handle different data formats, such as video, audio, or texts, while applying the inferencing techniques to edge nodes.
6. Edge accelerators: A few computing nodes support speeding up edge-level intelligence-related tasks. A few notable devices that promote acceleration involve TPUs, FPGAs, GPUs, etc. In this chapter, I explain the importance of such accelerators and highlight their placement in edge-level modeling or inferencing systems.
7. Performance/energy efficiency methods: Including edge nodes in an IoT-cloud application is crucial to improving the scalability and privacy of AI-based applications. In this chapter, I suggest the anticipated performance improvements and energy-efficient methods for such architectures. The energy efficiency techniques, such as changing the frequencies, including caching methods, modifying the compression levels, etc, are detailed in this chapter.
8. Edge-level security mechanisms: Securing data at the edge level for developing robust intelligent systems is one of the prime concerns of researchers and developers. In this chapter, I have detailed the security mechanisms that are developed for edge-intelligent frameworks or platforms. Additionally, in this chapter, a few core techniques that improve the privacy of end devices and objects are explained in detail with suitable architectures.
9. Edge AI frameworks and platforms: In the market, we can observe a few edge-AI frameworks or platforms. Some platforms are publicly hosted by cloud providers, and some other platforms are developed by academicians and researchers. There are several unique features on these platforms, therefore, in this chapter I have detailed the most commonly available edge AI frameworks for different domains and their advantages.
10. Orchestration platforms for edge–cloud continuum: Offloading tasks between the cloud and the edge requires automation. In this chapter, I have emphasized some of the key design considerations for orchestrating learning tasks. The chapter also suggests a few techniques that can be practiced to improve the quality of the learning processes of applications; in addition, it explains the platforms that extend such orchestrations, namely, AirFlow-based solutions.
11. Edge-based applications: Edge computing is applied in several application sectors, such as transportation, healthcare, smart cities, etc. In this chapter, I present some applications and ideas that augment edge intelligence. Most

emphasis has been given to a few notable animal emotion-related applications so that readers can infer some in-depth knowledge of one particular application domain.

12. Edge-enabled digital businesses: In general, there are tremendous possibilities to improve business due to IoT-enabled products. By including edge nodes, the economics involved in the applications benefits. In this chapter, therefore, I explain the economics involved in edge-based applications, and suggest cost improvements due to the inclusion of serverless-based edge solutions. Obviously, this chapter could encourage startups or entrepreneurs to utilize cutting-edge techniques associated with edge computing in their solutions.
13. Edge intelligence—challenges and future directions: In the last chapter of the book, I list a number of future outlooks while including edge-based intelligence in their applications or products. This chapter will provide novel ideas for promising researchers or practitioners in multiple domains.

References

[1] Awaisi K S, Abbas A, Khattak H A, Ahmad A, Ali M and Khalid A 2023 Deep reinforcement learning approach towards a smart parking architecture *Cluster Comput.* **26** 255–66

[2] Bellavista P, Penna R D, Foschini L and Scotece D 2020 Machine learning for predictive diagnostics at the edge: an IIoT practical example *ICC 2020—2020 IEEE Int. Conf. on Communications (ICC)* 1–7

[3] Benedict S 2022 Shared mobility intelligence using permissioned blockchains for smart cities *New Gener. Comput.* **40** 1009–27

[4] Chen Y, He S, Jin X, Wang Z, Wang F and Chen L 2023 Resource utilization and cost optimization oriented container placement for edge computing in industrial internet *J. Supercomput.* **79** 3821–49

[5] Guevara I, Ryan S, Singh A, Brandon C and Margaria T 2024 Edge IoT prototyping using model-driven representations: a use case for smart agriculture *Sensors* **24** 495

[6] Kannen N and Subasi A 2023 Smart factories of industry 4.0: determination of the effective smartphone position for human activity recognition using deep learning *Advanced Signal Processing for Industry 4.0* **vol 2** (Bristol: IOP Publishing) pp 3-1–24

[7] Lens.org 2024 Results The Lens–Free & Open Patent and Scholarly Search–lens.org (https://www.lens.org/lens/search/patent/list?preview=true&q=edge

[8] Lin D, Yao C, Min W, Han Q, He K, Yang Z, Lei X and Guo B 2024 A novel fall detection framework with age estimation based on cloud-fog computing architecture *IEEE Sens. J.* **24** 3058–71

[9] Pan J and McElhannon J 2018 Future edge cloud and edge computing for internet of things applications *IEEE Internet Things J.* **5** 439–49

[10] Qu Q, Liu C and Bao X 2021 E-commerce enterprise supply chain financing risk assessment based on linked data mining and edge computing *Mobile Inf. Syst.* **2021** 9938325

[11] Saranya T, Deisy C, Sridevi S and Anbananthen K S M 2023 A comparative study of deep learning and internet of things for precision agriculture *Eng. Appl. Artif. Intell.* **122** 106034

[12] Siddiqui S T, Khan M R, Khan Z, Rana N, Khan H and Alam Md I 2023 Significance of Internet-of-Things edge and fog computing in education sector *2023 International Conference on Smart Computing and Application (ICSCA)* 1–6

[13] Singh A and Chatterjee K 2023 Edge computing based secure health monitoring framework for electronic healthcare system *Cluster Comput.* **26** 1205–20

[14] Thakur D, Saini J K and Srinivasan S 2023 Deepthink IoT: the strength of deep learning in internet of things *Artif. Intell. Rev.* **56** 14663–730

[15] Urbaniak D, Rosell J and Suarez R 2022 Edge computing in autonomous and collaborative assembly lines *2022 IEEE 27th Int. Conf. on Emerging Technologies and Factory Automation (ETFA))* pp 1–4

IOP Publishing

Edge Intelligence
Deep learning-enabled edge computing
Shajulin Benedict

Chapter 2

Edge computing architectures

Before learning edge-level learning techniques and performance efficiency approaches for deep learning-enabled applications, it is essential to understand the available computing architectures that utilize edge devices as processing elements. We have to understand the basics of a few allied technologies that enrich the knowledge level of edge computing. The allied technologies that play a significant role in building up edge-enabled Internet of Things (IoT) applications are: (i) IoT, (ii) connection protocols, (iii) service-level protocols, and (iv) learning techniques, including deep learning. This chapter explains the base technologies that pave the way for readers to understand the edge computing architectures and associated learning methods.

2.1 Internet of Things

The Internet of Things (IoT) in a broad sense, is an extension of the normal internet. In the past, the internet was offered only to laptops, servers, or computational devices, including mobile phones. However, in recent years, the IoT has extended the internet not only to computational units but also to normal things and/or objects.

IoT devices are either lightweight devices such as pen, pencil, or rubber, heavyweight devices such as industry machines, tiny devices such as ear rings [4], capsules, or infrared robots, and large devices such as water quality sensors, wind direction sensors, or air quality sensors.

2.1.1 Subdomains of the IoT

In general, the IoT is subdivided into five broad categories depending on the functions or operational traits (see figure 2.1). The categories of the IoT and their key highlights are described below.

2.1.1.1 Industrial IoT

Industrial IoT is a revolutionary step for industries. Generally, industries improve their processes to generate the revenue of their operations. In fact, there have been

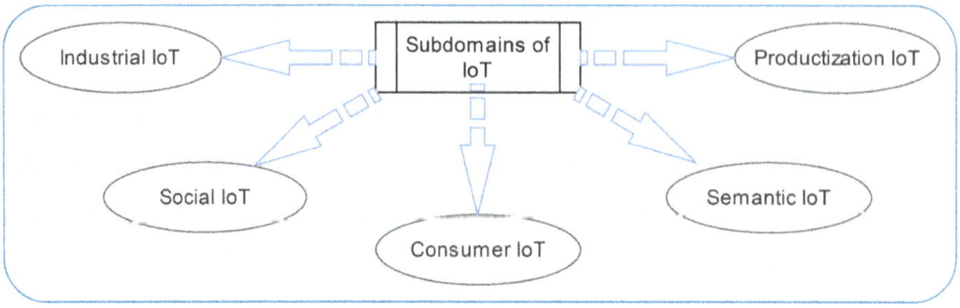

Figure 2.1. Subdomains of IoT technology.

several revolutions in industries in the past. Notably, the revolution due to steam engines in the eighteenth century increased the productivity of developing products. This revolution is termed the first revolution, or Industry 1.0. Next, in the beginning of the nineteenth century, electricity transformed the production processes in industries. It gave birth to Industry 2.0. This revolution increased the horsepower of engines and opened up opportunities to increase mass production in many existing solutions, i.e., improvement in production processes rigorously imposed cost efficiency in major industries, including automobiles. During this revolutionary period, employees of industries had to focus on a limited number of tasks when compared to the previous industrial periods due to mass-level production.

Next, in the early 1970s, the involvement of embedded microcontrollers and the application of computers paved the way for industries to operate machines in an automated manner. It also involved packaging messages in an electronic fashion to speed up decisions or share information. This revolution is termed Industry 3.0.

In Industry 4.0, industrial processes are enriched with networked things and objects in addition to traditional computers. Machines are connected with sensors and actuators to collect operational commands from subsequent machines. During this era, the digitalization of manufacturing products has been possible in an autonomous manner. Although a few more research projects need to be completed, the path toward reaching the goal of autonomous manufacturing is vividly appreciated by several industries.

2.1.1.2 Consumer IoT

In consumer IoT, personal devices are connected to satisfy the needs of users. For instance, devices such as laptops, wearables, voice assistants, robotic machines, smartphones, etc, can assist users in keeping them productive in their ongoing work. As an example, if a person wants to identify the best deal of the day on Amazon, they could ask or chat with the nearest device. Accordingly, these personal devices can check the best options based on the users' preferences and budget availability before placing an order in an autonomous fashion.

Consumer IoT aims to improve the comfort of users. In this way, users are able to focus on planned activities with major support from surrounding devices, mostly IoT-enabled ones.

2.1.1.3 Social IoT

In social IoT, social relationships are bestowed with predetermined preferences based on sensor-enabled technologies. Social IoT focuses on creating relationships among like-minded people, such as family members, friends, or objects related to connected people. In fact, all of the more than 7.5 billion people on Earth need relationships. Although we have relationships with family members, some of these relationships may be poor. Social IoT will have the potential to identify the best possible characters that match the huge volume of people around the globe.

For instance, a rich person may wish to find a poor person around the globe for various reasons, e.g., to help the needy person. Also, a politician may want to provide speeches to multiple communities without causing offence. Social IoT can provide the core preferences and interests of audiences so that the politician could utilize them to win hearts of many people.

2.1.1.4 Semantic IoT

Semantic IoT is another sub-division of IoT that is similar to social IoT. However, there is a slight difference between social IoT and semantic IoT. The social IoT is not focused on individual personal benefits. In contrast, semantic IoT is a tailor-made solution that maps the benefits of social IoT to personal need. For instance, if a teacher wants to identify the best explanation given to the students, semantic IoT could offer it. It applies ontologies and ontology-related web languages to process texts of a person based on their tailor-made characteristics.

2.1.1.5 Productization IoT

In productization IoT, sensor-based solutions are applied to design and manufacture products in a collaborative manner. The productization IoT approach improves the quality of products in a time-efficient manner due to the involvement of many developers in the design process, i.e., the productization IoT approach manufactures products in a collaborative manner while collecting feedback and design considerations from multiple volunteers. In fact, in the traditional approach to manufacturing a product, designs have to be reviewed by competent designers without sufficient rewards or penalties. In this method, feedback is collected and reviewed for efficient manufacturing synergies.

2.1.2 Service-level protocols

IoT devices require standards to efficiently communicate with other computing elements. In the IoT domain, there are tens of thousands of devices, mostly heterogeneous devices, that have to communicate with each other. In general, protocols are defined and standardized to facilitate efficient communications between multi-vendor or multi-characteristic devices.

Among these protocols, a few are utilized to establish signal-level communications; a few other protocols are designed for handling higher-layer communications, i.e., services, in an effective manner.

A few service-level protocols that are popularly applied in IoT-enabled devices are: (i) constrained application protocol (CoAP); (ii) message queue telemetry transport (MQTT); (iii) secured MQTT (SMQTT); (iv) extensible messaging and presence protocol (XMPP); and (iv) advanced message queue protocol (AMQP).

2.1.2.1 CoAP

CoAP [5] aims to promote web transfers on battery-operated sensor nodes. This protocol mainly focuses on machine-to-machine communications, similar to client–server communications. In the standard internet, messages are communicated or transferred in a client–server model. For instance, clients request information through browsers such as Firefox, Chrome, or Opera. These requests are processed by available servers, including Google's search engines, to deliver the responses. However, the traditional approach to receiving responses from servers is not an easy task for constrained devices. This is because of the heavy processing components involved in HTTP protocols.

CoAP serves just above the session layer of ISO/OSI networking standards. CoAP easily binds with HTTP protocols if the devices need to communicate with some powerful devices having sufficient power sources. The CoAP protocol has two different sublayers, which reside between the application and transport layers. These sublayers are named the req/res layer and the messaging layer.

1. Req/res layer is the request/response layer that handles the real communications of different constrained nodes.
2. Messaging layer, which resides between the transport layer and the req/res layer, is responsible for promoting reliability messages.

The advantages of the CoAP protocol are: (i) it provides reliability based on specific messages of the protocol; (ii) it suits well with HTTP protocols following TCP/IP standards; (iii) it can be easily integrated with HTTP-based communication devices; and (iv) it is energy-efficient as the protocol is operated nearer to the session layer, i.e., the functionalities in the top layers are minimized.

Although there are several advantages, a few disadvantages also exist in CoAP. They are: (i) the reliability messages are handled using UDP protocols in the transport layer. This reduces the chances of delivering UDPs to appropriate destinations. (ii) The design leads to an unordered delivery of reliability messages. Accordingly, there is a heavy computational requirement on servers associated with the protocol.

2.1.2.2 MQTT

MQTT protocol is a messaging protocol that follows the publish/subscribe model [3]. In the publish/subscribe model, devices can share data with intermediate brokers rather than waiting for responses from services. The traditional approach of sharing messages directly with listening services expects the servers and services to be online. However, in power-constrained environments, it is not advisable for the devices to be powered ON throughout the anticipated communications. Also, in IoT-enabled applications, the request for data is not frequent in many cases. Obviously, the request–response model will not fit very well in many situations.

In the MQTT protocol, intermediate brokers are required to satisfy as many sensors as possible. These brokers are involved in capturing sensor data from available sensors and delivering data to requested services. In general, these brokers have to be online throughout the communication period. But sensors and associated services need not wait for each other's availability.

The major steps involved in MQTT-based protocol communications are as follows:

1. Sensors publish sensor data to intermediate brokers. These sensors are not only hardware sensor devices such as temperature, humidity, pressure, etc, but also the data evolving from social IoT platforms.
2. Services connect with intermediate brokers for subscribing to sensor data. These services include visualization services on mobile phones, alarm and notification services on cloud nodes, etc. These services, when hosted on cloud servers, are liable to pay the required costs associated with them.

2.1.2.3 Secured MQTT

The secured MQTT protocol is similar to the secured HTTP protocol. The secured MQTT protocol applies the publish–subscribe model, i.e., it allows sensor nodes to publish sensor data based on the availability of sensors, and it enables services to subscribe the sensor data through intermediate brokers in a secured manner, i.e., the secured MQTT includes some cryptographic algorithms to encrypt data between communicating devices. These security-based algorithms can draw battery power if they are not handled in an efficient manner.

Researchers in [2] have studied the performance of encrypting messages between MQTT clients and servers. It can be observed that certain applications have to compromise a few CPU cycles in order to achieve better-secured communications.

2.1.2.4 XMPP

XMPP is based on the XML language that conveys communicating messages over the SOAP protocol. The protocol focuses on instant messaging services. Traditionally, XMPP was designed for server-based applications. However, in the later stages, considering the resource-constrained features of IoT applications, researchers proposed a few lightweight versions of them. Notably, the authors of [17] developed a lightweight XMPP protocol that has the capability to downsize the traditional XMPP protocol by tuning the sleeping mechanisms of algorithms.

The lightweight versions of XMPP protocols paved the way for improving the utility of the protocol in several IoT-enabled applications, including edge-AI. For instance, if sensor data needs to be transferred quickly between connected devices or edge-AI services, then XMPP would be a candidate protocol at the service level. Basically, XMPP routes XML data in near real-time in a structured data format. There are a few XMPP-based protocols: they are Core, Jiggle, PubSub, BOSH, etc. The major drawbacks of XMPP protocols are non-available QoS services and potential applications.

2.1.2.5 AMQP

In the AMQP, messages are routed while following the publish–subscribe model. To do so, brokers are extended to involve routing mechanisms, i.e., specific routing approaches are specified in the brokers so that security is ensured between sensors and associated services.

If publishers are not configured properly, there is a higher probability of intruding on exposed open brokers. Consequently, machines could damage devices or take advantage of controlling other machines while performing operations. Hence, AMQP protocols, with the engagement of routing processes, will support solving the security issues. The AMQP protocol is applied to modern brokers such as RabbitMQ. It can be seen that researchers have utilized AMQP protocols to improve the productivity of industries. [7].

Figure 2.2 illustrates the taxonomy of service-level communication protocols that are utilized by edge-level computing nodes to connect them with sensors and actuators.

2.1.3 Communication-level protocols

A few communication protocols that are utilized by sensors to connect with other devices are described in the following sub-sections. Learning these communication protocols enables readers to understand the applicability of appropriate IoT-based communication protocols for minimizing the data communications or network latency issues.

2.1.3.1 IEEE 802.15.4 standard

The majority of the sensor nodes follow IEEE 802.15.4 standards owing to battery limitations. In general, IEEE 802.15.4 is a standard defined for a low-rate personal

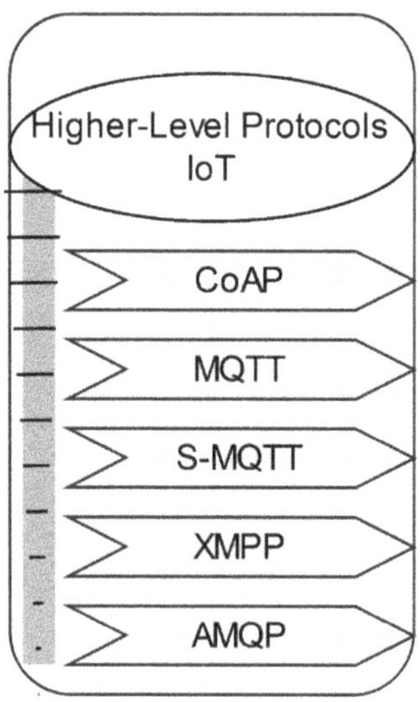

Figure 2.2. Taxonomy of service-level communication protocols in IoT.

area network that reduces the battery utilization or energy consumption of applications. The devices that operate under this standard have a 2.4 GHz operational frequency. Also, the devices apply spread spectrum modulation techniques rather than heavyweight modulation techniques.

The devices that follow the IEEE 802.15.4 standard have two sublayers in the data link layer. These two sublayers are named the service-specific convergence sublayer and the logical link control layer.

The medium of access is dependent on beacons. These devices follow beacon-enabled methods, considering the carrier's multiple access policies. This means that beacons are transmitted periodically to understand the availability of nodes in a transmission channel. Depending on the responses received from devices in a special packet format, the communication medium is allotted.

In such devices, power consumption is reduced from 0.5 mW to 1 mW. This paves the way for the devices to have increased battery life. Additionally, the transmission range is often 40 m in outdoor communications where line of sight (LoS) is maintained. However, the preferred distance between communicating devices is limited to 10 m, especially when the devices are operated within buildings.

2.1.3.2 Bluetooth

The Bluetooth protocol is popular on low-power devices to enable secure communications using an unlicensed ISM band, i.e., 2.4 GHz. Compared to traditional versions of Bluetooth, the devices following the Bluetooth protocol could support seven more devices at a data rate of 1 to 3 Mbps. The protocol enables devices to cover a limited range of 10 meters using frequency-hopping spread spectrum (FHSS) modulation techniques.

The Bluetooth protocol operates by forming a small network named piconets, i.e., devices connect and communicate with each other using a smaller number of networking nodes. The operations of seven other devices follow the master–slave approach. As the name indicates, one device operates as the master, while the other connected nodes in a piconet serve as the master node.

The Bluetooth protocol has evolved over time. Starting with Bluetooth 1.0, the features varied in architecture and connection approaches to acccommodate the requirements of energy efficiency and communication quality. Recently, Bluetooth 5.1 was released. This version has improved the periodic advertisement interval and the encryption techniques.

2.1.3.3 LoRaWAN

LoRaWAN is a long-range wide-area network. As the name indicates, this protocol enables wide-area connectivity ranging from 500 meters in urban areas to 10 kilometers in rural areas where LoS exists between the transmitter and receivers.

The idea of increasing the range of communications is based on minimizing the bandwidth of communications. LoRaWAN promotes multiple ranges of frequencies. For instance, 868 MHz is utilized in Europe, 915 MHz is applied in North America, and 433 MHz is used in Asia. Recently, the authors of [10] have developed a multi-hop LoRaWAN protocol that survives a long range of communication between tiny sensor nodes to enable the required data analysis.

Most LoRaWAN devices, either sensors or actuators, are often tiny battery-operated devices. These devices send short messages such as the value of temperature, humidity, pressure, etc. LoRaWAN utilizes the concept of a point-to-point, multi-point communication approach to handle a large volume of sensor nodes within a wide area. The protocol is most commonly applied in industries to enable machine-to-machine communications. For instance, a machine submits its status to the subsequent machines to continue the workflow without any human intervention. In such situations, LoRaWAN gateways are mounted on a factory site comprising over 1 hectare of land.

2.1.3.4 WiFi

WiFi is a wireless fidelity concept that boosts common features between different manufacturers. Wireless devices, manufactured by multiple vendors, are supposed to follow IEEE 802.11 wireless standards. The governing or regulating body named the WiFi Alliance is responsible for formulating newer definitions of standards. WiFi, in general, provides wireless LAN connections.

WiFi-enabled devices that follow the IEEE 802.11 standard have the capability to form ad hoc networks apart from traditional infrastructure-oriented communications. WiFi has several implementation versions, such as v3, v4, v5, and v6. There are multiple standards depending on the speed of communications. Notably, IEEE 802.11 started standard communications in the range of 2 Mbps. Next, IEEE 802.11b increased the data rate to 11 Mbps. Next, IEEE 802.11a/g was designed to support a data rate of 54 Mbps. The recent data rate of WiFi has been increased to 1.7 Gbps and 2.4 Gbps while considering IEEE standards such as IEEE 802.11ac and IEEE 802.11ax. Although the majority of the standards support the ISM band, the recent WiFi communication systems are designed to handle 5 GHz bandwidth in a multiple input, multiple output MIMO mode.

2.1.3.5 5G/6G mobile networks

Wireless communication through mobile networks or cellular networks is a peculiar approach in next-generation communication systems. Cellular networks have seen several generations of technologies over a period of time. For instance, 1G, 2G, 3G, 4G, 5G, and 6G have improved the speed of communications and support for multiple input features.

In fact, the first-generation mobile network was involved in analog communications. This generation utilized circuit switching networks rather than packet switching. The first generation of mobile networks has a speed of 2.4 kbps with high latency, i.e., a latency of over 1000 milliseconds. Obviously, it is not a reliable communication method.

In second-generation systems, the mobile network was modified for digital communications. In this generation, circuit switching concepts were replaced with packet switching communications. Owing to the digital communication approach, voice compression techniques were incorporated into the solutions. This led to more reliable communications with better latency. In 2G mobile networks, the latency was typically around 1000 milliseconds. And the speed of communications reached around 64 kbps. Accordingly, multimedia messages and short messages were popular in such systems.

In third-generation mobile networks, data communication was introduced apart from traditional voice signals. In this era, the data rate was increased to 2 Mbps, which increased the potential communication capabilities such as video-based applications, audio conferencing, gaming, etc. In fact, before the start of 3G, there were 2.5G and enhanced data rates for GSM evolution (EDGE) systems. The 2.5G introduced the concept of general packet radio services (GPRS). It was a very important feature for several GSM users. The introduction of EDGE systems reduced the latency of applications to less than 1000 milliseconds. Similar to 2.5G and EDGE, the third generation had a few intermediate steps before entering 4G. They are referred to as 3G and 3.5G. The third generation introduced the CDMA medium access technique with an increased data rate of around 14 Mbps in 3.5G and 42 Mbps in 3.75G. In most mobile phones, we could observe the indication of 3.75G as H+ symbols.

The demand for speed obviously doesn't stop among mobile users, as the majority of the users are keen to pursue novel applications. The birth of 4G, in fact, was a step to meet most of the demands of trending mobile users. It enabled them to have online classes, online video conferencing, video gaming, etc. This generation delivered around 1 Gbps of data to offer realistic HD video-related applications. It has been a reality in many cities since 2019.

Currently, the 5G market has started to penetrate the minds of mobile users as 4G is challenged in higher latencies. Notably, 4G has experienced over 200 milliseconds of latency that have delayed real-world applications, including real-time operations and virtual reality.

In 5G, the latency of applications has dropped to less than 1 millisecond. Obviously, any real-time application is possible with this approach. Autonomous vehicles could brake in almost real time based on the commands that are remotely submitted to braking systems. Similarly, doctors could realistically perform remote operations due to the minimized latency factors. Several research works have been undertaken in the recent past to understand the latency improvements due to 5G or later generations in mobile networks [9].

5G is operated in three major bands. They are termed (i) high band, (ii) mid band, and (iii) low band. In the high-band spectrum, we experience a true 5G system owing to the higher bandwidth, i.e., 25 to 39 GHz. This spectrum is also named mmWave (millimeter wave). The data rate of devices operating in this spectrum is around 3 to 10 Gbps. This is almost 10 times the speed of its successor, i.e., 4G networks. However, it has limited coverage, which extends from 10 to 50 meters.

In mid band, 5G spectrum is classified into two subcategories: (a) mid band–I and (b) mid band–II. The two mid bands are designed for 5G in order to reach a long range, a short range, or a long distance. This band is quite suitable for smart cities and towns that reach a distance of 10 to 30 km. In this band, the devices operate at 2.5 to 4.2 GHz with a speed of 100 Mbps to 1 Gbps. In this band, the users experience a speed of 4G mobile connections. Since mid band has comparatively fewer bands, it could reach larger distances when compared to the high-band 5G spectrum.

The low band 5G spectrum is aimed at increasing the coverage of mobile devices. To do so, a low band is utilized in the range of 600 to 850 MHz. Obviously, 5G devices could cover hundreds of square miles. This band will not provide a true 5G experience. Hence, it is often referred to as a false 5G.

2.1.3.6 5G massive MIMO

MIMO, in general, resembles multiple input multiple output (MIMO) in transmission systems. In fact, it is an antenna expansion concept that attempts to increase the number of antennas and receivers. Using MIMO, the transmission systems capture the multipath signals using low-energy transmissions. The idea of MIMO proliferated prior to 5G. However, in 4G technologies, the number of antennas and receivers is limited. Typically, two antennas are utilized in such technologies.

In 5G, massive MIMO is incorporated. In this way, many transmitters and receivers are capable of directly connecting to each other. The base stations are able to point toward mobile phones or gadgets during the call establishments.

Due to the inclusion of massive MIMO, 5G is capable of increasing spectral efficiency by satisfying multiple users. The antenna technology improves by forming array antennas with lower power transmission. The antenna multiplexing gain and antenna array gain have dramatically increased with the massive MIMO technology.

2.1.3.7 5G beam forming

5G supports beam forming to accommodate more users at a higher data rate. It targets the direction and movement of receiving devices with multiple radiations simultaneously from an array of transmitting antennas. The antennas of a 5G massive MIMO are not heavy, i.e., a device weighs around 25 kg.

Although beam forming directs the waves toward receiving mobile phones, the computations involved in the process are much more complicated. The directions need to be assessed in real time while the receptors move within the vicinity of the antenna or within the vicinity of adjacent antennas. Obviously, some artificial intelligence (AI) techniques are applied to predict the requirements of beam forming for a particular device.

2.1.3.8 5G network slicing

Network slicing enables the virtualization of network infrastructures in a lightweight fashion. Instead of holding multiple physical infrastructures with different configurational settings, it is possible to create virtual instances of such devices. The approach eases network administrators as multiple configurations are carried out using software-driven methods.

In 5G, end-to-end network slicing is developed. This means a logical network is provided to end customers. With the development of slices, users could experience different networks that are customized to their businesses. Slicing is performed based on software-defined networks (SDNs) and network function virtualization (NFV).

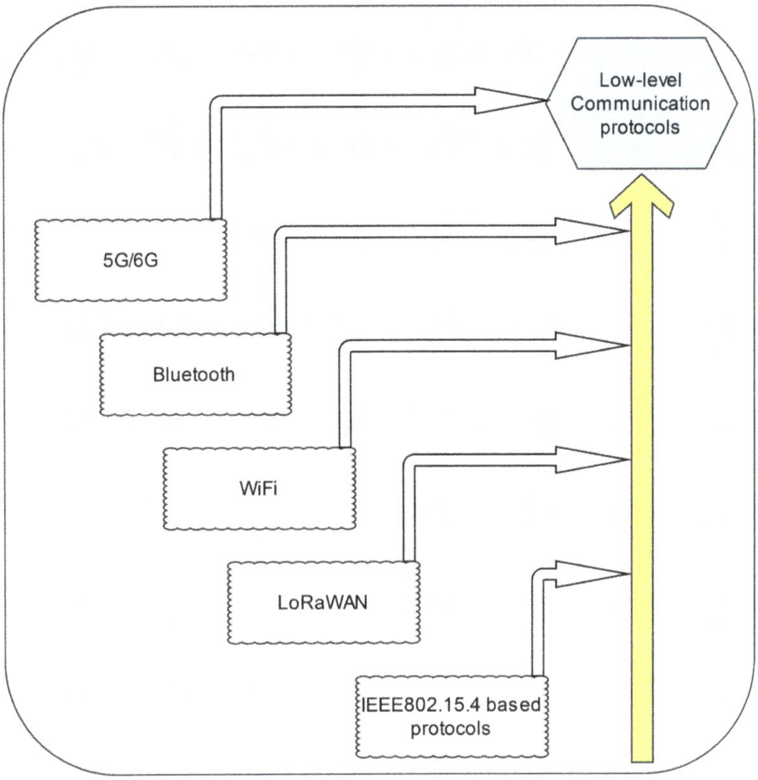

Figure 2.3. Taxonomy of low-level communication protocols in IoT.

In NFV, network functions such as framing policies, firewall features, routing approaches, etc, are virtualized in the form of containerized solutions that satisfy the specific network requirements of users.

As there are several parameters to decide the network slicing components in 5G-enabled devices, the authors of [16] developed a reinforcement learning algorithm-assisted network slicing mechanism.

Figure 2.3 illustrates the taxonomy of lower-level communication protocols that are utilized by edge-level computing nodes to connect with sensors and actuators.

2.2 IoT-enabled components

IoT-enabled components comprise devices belonging to various computational capacities and administrative controls. An IoT-enabled ecosystem, in general, has sensors, edge/fog nodes, cloud servers, user interfaces, and a few software components, including analytics.

2.2.1 Sensors

There are tens of thousands of sensors that are utilized in any scalable IoT application. For instance, an IoT application that captures the health status of people living in a smart city can contain millions of sensors. These sensors might belong to different vendors and have different characteristics.

2.2.1.1 Sensor characteristics

Sensors are characterized based on their sensing capabilities. The majority of the sensors are accurate in producing results. However, there are sensors that don't require 100% accuracy while sensing properties. Sensors, having the capability of measuring properties such as light, temperature, humidity, etc, must measure them with the utmost accuracy. Also, these devices are expected to produce the same results when repeatedly experimented with measuring properties. In some cases, the sensors are supposed to drift away from the accuracy of the production of sensing values. However, they may be linearly drifting away from the normal values or nonlinear.

2.2.1.2 Sensor classification

Sensors can be classified in several ways based on their size, measuring properties, activeness, scalar/vector, and analog/digital.

1. Sensor size: They are classified depending on the size of the sensors. For instance, sensors can be tiny or large based on their measuring properties. For instance, a temperature sensor is smaller in size, ranging in the order of millimeters. However, a rain sensor is larger in size, ranging in the order of centimeters. There are also huge sensors in the range of meters to measure weather conditions. Obviously, it is easy to observe that sensors vary in terms of their size.
2. Measuring properties: Sensors are classified based on their measurement properties, such as temperature, humidity, chemicals, gas, etc.
3. Activeness: We could classify sensors based on their activeness. For instance, some sensors are able to capture data without the need for any power sources. These types of sensors are termed active sensors. In contrast, sensors that require external power sources to sense data or measurable properties are termed passive sensors.
4. Scalar/vector: Sensors can be classified based on the scalar or vector properties of devices. For instance, sensors that measure properties without indicating directions are termed scalar sensors, and sensors that measure properties with values that include scalar components and directions are termed vectors. Some typical examples of scalar sensors are temperature sensors, ECG sensors, water quality sensors, etc. Examples of vector sensors are camera sensors and unidirectional audio sensors.
5. Analog/digital: Sensors are capable of collecting digital or discrete values, for example, digital cameras. In some examples, sensors measure continuous, measurable properties. The most commonly utilized analog sensor is the temperature sensor.

2.2.2 Edge devices

Edge devices are hardware devices that are meant to control the flow of information between sensors and clouds using distributed architectures. They are often connected to sensors or nearer to sensors to process sensor data. These processing steps involve filtering information, averaging, etc.

2.2.2.1 Properties of edge devices

Edge devices have special characteristics that uniquely distinguish them among other nodes. The most notable characteristics are listed below, along with some descriptions:

1. Support for communication protocols: IoT-enabled devices are manufactured to emit different types of communications, such as WiFi, Bluetooth, Zigbee, LoRAWAN, etc. The edge devices that are designed to support these varieties of communication protocols should have the capabilities to support them. The edge-based hardware should be manufactured to accept the majority of these communication protocols.
2. Internet capability: Edge devices should have IP addresses to uniquely identify the device. In some cases, these devices should have the support to deliver domain names for underlying connected sensor nodes.
3. Variable speeds: Edge devices must have the capabilities to buffer data to support the variable bandwidths of connected networks. The networks connected to upstream scalable devices or cloud networks will have higher speed requirements when compared to the sensor nodes.
4. Cost-efficient parallelism: Edge devices are often considered to be less expensive than heavyweight computational nodes. In recent years, researchers have attempted to cluster an array of low-expense microcontrollers, such as Espressif's ESP nodes, to envision massive parallelism. For instance, the authors of [11] have demonstrated the execution of federated learning models using edge nodes applying parallelism.
5. Fanless support: Edge nodes are supposed to not require fans for cooling processors. This is because, on the majority of devices, edge nodes stop functioning due to the failure of fans. If edge devices are designed without fans, the noise level is minimized, and they can be mounted in remote locations where maintenance needs are minimized [15].
6. Secure environments: Edge devices must have sufficient security features to protect devices and solutions. In fact, IoT-enabled solutions must consider the battery limitations before including all relevant security assistance such as authentication, authorization, encryption, integrity checks, etc.

2.2.2.2 Types of edge nodes

In any IoT-enabled application, intelligence starts with edge nodes. Sensors are not capable of performing intelligent tasks. They are meant to measure a physical property such as temperature, humidity, pressure, etc. However, they could perform intelligence while combining the tasks with edge nodes. Some edge nodes are faster than widely applied edge devices.

Edge nodes are classified based on various factors: (i) processing capabilities and hardware components, and (ii) deployment locations.

1. Processing capabilities and hardware components: Edge devices have different processing capabilities. Some edge devices have one CPU, and others have multiple CPUs. In some cases, devices include GPUs, while others include TPUs or FPGAs. Based on the associated computing devices, the processing capabilities of these devices vary. For instance, multi-core CPUs have 4, 8, 16, 32, 64, or 128 CPUs to execute tasks in parallel. These CPU cores are hardware components that compute sets of instructions with their own registers.

   However, GPUs are graphic processors that act as co-processors to speed up data-intensive instructions. Notably, instructions that have multiple different input values are processed on GPU-based devices. These computing components are not able to execute the whole process without the support of CPUs.

   In recent years, due to the engagement of several machine learning algorithms, a few application-specific integrated circuit (ASIC) devices have been developed. ASICs are particularly designed for accomplishing the goals of applications. For instance, an ASIC designed for deep learning algorithms will have better performance than the traditionally available hardware devices while executing applications such as object detection, object segmentation, object classification, and image analysis. The authors of [12] studied the performance efficiency of different edge-level hardware architectures.

   Field-programmable gate arrays (FPGAs) are uniquely designed hardware that has provisions to modify hardware circuits for different applications. Irrespective of software modifications, in some cases we have to fine-tune the characteristics of hardware components. In this way, parallel executions are faster without the involvement of central processors.

   For instance, assume that we have to design a microcontroller device that has a few connected actuators. Let these actuators act depending on the sensory information received from sensors. In the traditional case, microcontrollers have to process the input data from these sensors and decide on particular actions that the actuators must execute. However, we could bypass this process with the support of FPGAs. FPGAs are hardware-level integrated solutions that bring forth the idea of establishing configurable logic blocks for applications. Obviously, the performance of such applications increases. For instance, the authors of [19] have demonstrated the application of FPGAs for object detection-based deep learning algorithms.

   The major drawbacks of FPGAs are: (i) programming at the hardware level is required; (ii) skilled workers or engineers are required to operate the devices; (iii) they are limited to specific applications; and (iv) they are expensive when other software-related solutions are considered.
2. Deployment locations: Edge concepts are deployed in several locations in a decentralized computing system. Depending on the locations of deployment, we could classify edge nodes into the following categories: (i) far-edge level, and (ii) near-edge level.

(a) Far-edge-level processing is carried out on sensors or next to them. For example, sensors such as camera sensors and wind-data processing sensors have streaming data. These streams are either periodic or aperiodic in nature. As a large volume of data needs to be transferred to higher computational nodes, camera sensors or wind-data processing sensors have pre-built computing units along with the sensors. The objectives of these sensors are to pre-process the data, filter the data, or undertake data-level operations such as color removal, image enhancement, zooming, etc.

Based on the availability of power sources, the edge nodes at this level are described as battery-sourced and power-sourced edge nodes. As the name indicates, some camera sensors are able to operate using batteries in remote locations. These sensors involve far-edge-level processing sourced through battery power. The advantages of such devices are: (i) they are able to reside in any remote location; (ii) they are often lightweight and tiny devices. However, the disadvantage of these types of devices is that they need the batteries to be replaced very often. Obviously, we could not place them in inaccessible locations. In contrast, the power-sourced nodes are often capable of handling relatively large operations without concern for their power limitations. The far-edge devices are most similar to smart sensors or intelligent gateways that have some computational capabilities to process sensor data.

(b) Near-edge-level processing includes a powerful server for performing edge-level capabilities. These servers are placed far away from sensors. They scale well when compared to far-edge nodes, as they are closer to clouds. The major difference between such devices and clouds is that they are often battery-operated and have fanless support to integrate well with scalable environments. Particularly, edge nodes operated in the form of server nodes have more compute capabilities than edge capacities housed on sensors. Also, deploying near-edge-level functions on server nodes has the additional advantage of incorporating multiple communication channels such as LoRaWAN, Zigbee, WiFi, Bluetooth, etc.

The major objectives of including edge nodes on IoT-enabled applications are: (i) to increase the performance of applications; (ii) to reduce the latency of applications, especially when they are data-intensive in nature; (iii) to enrich the privacy and security concerns of users, i.e., data resides nearer to the user; (iv) to minimize the network bandwidth of the entire IoT-enabled application; (v) to efficiently handle the data storage units of applications: i.e., storing data on cloud storage units is minimized etc.

2.2.3 Fog nodes

An IoT-enabled application involves not only an edge-cloud combination of computing nodes but also fog nodes in a few cases. In edge computing nodes, there are limitations in storing data owing to minimal memory space. The involvement of fog nodes is mainly focused on designating storage units.

2.2.3.1 Fog: a decentralized cluster

In general, fog computing involves clusters that consist of a larger number of servers that are connected using high-speed network connections. These are powerful computing units often deployed in city centers or factory units with more storage capabilities. These cluster units reside between centralized cloud computing data centers and edge nodes. Fog nodes are highly decentralized computing nodes and storage nodes that have the capabilities to connect multiple edge nodes; they are also capable of executing modeling algorithms, including reinforcement learning algorithms, by deploying multiple agents across the network.

2.2.3.2 Fog: characteristics

Fog nodes are uniquely characterized depending on their applications, as follows:

1. Low-latency: Compared to cloud nodes, fog can deliver low latency to IoT-enabled applications. For instance, a healthcare application streams ECG data to cloud services to learn the ECG pattern of a patient. In general, edge nodes cannot learn the history of patients as they are often not connected with the medical records of hospitals. Developing and deploying learning algorithms on clouds, however, increases the latency of applications. Hence, fog clusters could ease the purpose of applications.
2. Location-based: Servers are clustered in city centers or at specific locations in fog-based establishments. Obviously, these locations house the required data for processing them. For instance, hospital records can reside within their network jurisdiction or administration. Hence, data-centric databases could be designed for fog-enabled healthcare applications.
3. Real-time: Fog nodes attempt to deliver a real-time data analysis solution. For instance, enabling augmented reality while designing a new product on a manufacturing platform requires real-time processing of data; performing operations by a group of doctors is one of the real-time use cases that requires robust computing devices such as fog nodes nearer to the sensor data.
4. Security: Fogs are implemented by city authorities or industrialists. Such fog-based setups are administered by the end user's administration. For instance, a fog cluster installed in a smart city offers fog-enabled solutions in its city; hospitals deploy fog-enabled compute clusters within the group of hospitals they belong to.

2.2.3.3 Advantages and disadvantages of fogs

The major advantages of involving fog clusters are: (i) extended security when compared to clouds: data is processed nearer to the sensors; (ii) low latency; (iii) increased bandwidth; (iv) engagement of more applications etc.

However, there are a few disadvantages to utilizing fog nodes [1]. They are: (i) *Trust issues*: fog is still considered a mini-cloud environment. Hence, there is a challenge in trusting providers. For instance, there are multiple government departments that cannot directly trust others. For instance, a corrupt police force could manipulate investigation-related data stored in fog clusters. This vividly demonstrates that fog clusters can raise trust-related issues. (ii) *Configurational issues*: Fogs are not well maintained, similar to clouds. Skilled technicians are required in fog cluster sites to solve any configurational errors or maintenance-related issues. Automatic solutions could also hamper the functions of fogs. Obviously, delayed downtime is possible due to the non-availability of skilled laborers.

2.2.4 Cloud nodes

It is important to understand the difference between cloud and edge nodes. In a broad sense, clouds are massive devices hosted by third-party providers such as Amazon AWS, Google Inc., Microsoft, Alibaba, etc.

2.2.4.1 Cloud services

The cloud environments offer three important services: infrastructure-as-a-service (IaaS), platform-as-a-service (PaaS), and software-as-a-service (SaaS).

1. IaaS: IaaS offers compute, storage, or networking-related components as a service. Offering hardware as a service has created a revolution in the computing space. Prior to this, resources were shared based on known connections or organizational policies. For instance, a researcher had to write a detailed proposal specifying what kind of resources are required from the provider, often supercomputers from academic organizations. This approach was not feasible for many common researchers or practitioners, including students. IaaS provided all hardware resources based on payments. However, IaaS expects users to install relevant platforms and software for executing them.
2. PaaS: In PaaS, the platform is offered as a service so that users can directly install their application or software. In fact, PaaS is very beneficial to startups or entrepreneurs, as they can focus on developing their own logic instead of spending manpower on the underlying platforms For instance, a deep learning-based startup needs to develop and install execution environments such as TensorFlow before focusing on designing new deep learning algorithms. The PaaS services, however, require limited knowledge of hardware or associated implementation bottlenecks. In fact, the majority of issues happen due to hardware-related issues, especially when networking aspects need to be dealt with.
3. SaaS: The third service in cloud setup is the implementation of software-as-a-service. Here, the whole software stack, along with the required platform and infrastructure, is offered as a service to users. In this approach, users may have to focus on feeding data rather than writing new algorithms. SaaS is useful for end users who prefer to execute business logic that is framed by providers. SaaS bestows less flexibility on users for updating software or concepts.

2.2.4.2 Cloud deployment models

Typically, cloud services are deployed in four ways. Depending on the deployment model, clouds could be categorized as: (a) private clouds, (b) public clouds, (c) hybrid clouds, and (d) community clouds. The services are deployed in a varied fashion to associate resources with different users. The most crucial characteristics of these different deployment models are described below:

1. Private cloud: A private cloud is a deployment model where resources are shared by a single organization among its customers or users. For instance, an academic organization could share compute or storage resources as services such as IaaS, PaaS, or SaaS with their stakeholders. Students, faculties, or staff need not enter into strict policies as resources are within their own organization.

   Although security is higher in a private cloud deployment model, the cost of the infrastructure is very high as resources need to be purchased and maintained within an organization.
2. Public cloud: In the public cloud deployment model, resources are hosted by third-party providers. The most popular third-party providers include Amazon, Google, Microsoft, etc. Third-party providers are often companies that dedicate expensive resources, such as data centers. These companies will have their own networking support procedures and manpower to assist with the requirements of users or customers.

   The most important advantages of public clouds are: (i) highly scalable resource availability; (ii) zero maintenance for customers and users; (iii) reasonable security; (iv) fast update or upgrade options, including hardware resources; and (v) higher resource utilization.

   However, public clouds undergo vendor lock-in issues in many situations. In vendor lock-in problems, customers are either directly or indirectly locked to their services. In this way, the users are forced to utilize the public cloud services without having special improvements.
3. Hybrid cloud: The hybrid cloud deployment model allows the combination of both private and public cloud infrastructures. This deployment model gives private cloud providers more flexibility to offload their resources. For instance, an organization could set up a private cloud with available resources. However, to satisfy the demanding requirements, certain resources may be rented from public clouds.

   In order to increase the security objectives of clouds, in most deployments, practitioners attempt to provide private cloud infrastructures when security-conscious workloads need to be handled, and they offer tasks that are not very critical and are submitted to clouds. As hybrid cloud deployment models combine the features of private and public clouds, they gain the most important features of them, i.e., scalability, resource optimization, security, and cost effectiveness.
4. Community cloud: This deployment model involves the specialized deployment of resources among communities. Communities such as drug discovery,

science, galaxy study, etc, might require private infrastructure from multiple organizations. For instance, forensic departments from multiple countries might share data or equipment to quickly identify international criminals.

The deployment options of the cloud could be differentiated based on providers, size of subscribers or services, security capabilities, scalability capabilities, managerial overheads, operational costs, and target customers.

2.3 Learning techniques: machine learning

Learning is an important task for existence. Humans learn a lot of information from childhood due to the input of trained information from mothers, school teachers, experienced faculties, experts, etc. The learning processes in humans have been modeled in several ways in machines so that the machines could learn and infer data from the environment.

2.3.1 Machine learning: the history

In fact, the learning approach was initially started in the form of AI in the early 1950s for machines. During this period, rules and logic have been the most important ingredients in learning algorithms that are executed on machines. Unfortunately, the experts have to be more vigilant about framing the rules and logics so that the learning output is reasonably attractive to the users. Although the process was quite acceptable to an array of researchers during this time, the procedure was not well suited to many working environments as there was a need for people who could evaluate the logic and rules that were customized to the situations.

In the early 2000s, the concept of machine learning evolved due to the difficulties involved in traditional AI-based reasoning or learning processes. In machine learning, the data and outcomes are provided instead of rules or logic for learning an environment. The outcome of machine learning is the formation of a set of rules and logic so that new findings and learnings are more realistic in many working environments.

Accordingly, several learning algorithms were developed based on machine learning concepts so that newer logics are framed by the algorithms given the input data and environmental outcomes. For instance, algorithms such as random forests, support vector machines, XGBoost, perceptrons, etc, evolved at a faster pace across various domains.

2.3.2 Algorithmic views

Within a short span of time, the growth of machine learning was so prominent that it was classified based on three types: (i) supervised learning, (ii) unsupervised learning, and (iii) reinforcement learning. These three types are described below:

1. Supervised learning: Supervised learning is a learning process in which the targets are reasoned based on labeled datasets: i.e., the training dataset. The labels are suggested by experts or experienced systems. Therefore, the data

and the outcome of the inferences are known in advance for the training data in supervised learning. Accordingly, the new inferences are formulated depending on the new dataset. A few algorithms that are categorized under the supervised learning process are random forests, regression algorithms, and support vector machines.

2. Unsupervised learning: In unsupervised learning, the data is not labeled prior to the learning process. It is beneficial when the data is not known by learning algorithms or when the features of the given data are not engineered. Although the data need not be labeled, this learning process suffers from deriving knowledge from data owing to the complexity involved in feature engineering. Also, there is no specified feedback for algorithms for iterative learning. Hence, unsupervised learning processes are more suitable for deep learning algorithms than machine learning algorithms. A few algorithms that fall under the unsupervised learning algorithm category include the *a priori* algorithm, k-means clustering, and hierarchical clustering algorithms.
3. Reinforcement learning: Reinforcement learning is a subset of learning in which the learning is dependent on the environment's optimal conditions. This learning process works well when the environment is not known to the learning algorithm. The learning algorithm has to study the environment through rewards or punishments acquired over time. Typically, a few learning agents are involved in these processes to acquire knowledge from the environment and suggest positive or negative decisions.

2.4 Learning techniques: deep learning

Deep learning is a sub-field of AI and machine learning. The idea of deep learning is to delve deeper into data to understand the features and representations so that the decisions are more accurate than machine learning algorithms. The major differentiation of deep learning when compared to machine learning is the type of data that is utilized in the deep learning processes. The deep learning algorithms can consider unstructured data while learning it. In contrast, machine learning algorithms expect structured data.

Deep learning includes several layers of learning representations from input data. The input data could be images, videos, audio, etc. The image input passes through several learning layers that convert the image to meaningful representations across layers. The idea behind having such learning layers is to fine-tune the representations of data to attain a more meaningful representation from the final layer of the deep learning algorithm.

Usually, deep learning algorithms are classified into several categories depending on the input data and learning requirements. The most commonly utilized types of deep learning algorithms are as follows:

1. Multi-layer perceptron: The multi-layer perceptron is a deep learning network that consists of multiple input layers, output layers, and hidden layers. The layers are fully connected and take vector inputs. The perceptrons follow

a feed-forward neural network approach, which is a good fit for single image classification problems.

2. Convolutional neural network (CNN): CNNs are the most widely utilized deep learning algorithms. They consist of convolution layers, activation layers, pooling layers, and fully connected layers for classifying complex images and input datasets. They are applied in several computer vision-related tasks as they can detect the spatial hierarchical features of datasets.
3. Object detection algorithms: There are a few deep learning algorithms, such as recurrent CNN, fast R-CNN, faster R-CNN, YOLO, and single shot detectors (SSDs) for detecting objects from images or video frames. These deep learning algorithms have their own uniqueness in detecting complex images, either iteratively or in a single stretch.
4. Recurrent neural network (RNN): This is a type of deep learning algorithm that fits well with sequential data. The algorithm has unique characteristics for detecting temporal features of data in a time-series manner. Obviously, there are several applications that could utilize RNNs to perform deep learning functions. For instance, memorizing old sequences of data in RNNs improves learning experiences in a subset of applications such as natural language processing or word prediction problems.
5. Autoencoders: In autoencoding, data encoding methods are applied to compress input data by extracting features of the data: i.e., the conversion of high-dimensional input data to low-dimensional data. In general, autoencoders are preferred when data and learning models have to be processed on edge-enabled, resource-constrained devices. It is a learning algorithm that falls under the unsupervised learning algorithmic category. Variational autoencoders are a subcategory of autoencoders.
6. Generative adversarial networks (GANs): GANs are a type of deep learning model that could generate datasets based on specific features [6]. For instance, generating human faces relating to animal faces, creating funny cartoons, formulating newer designs that are specific to application domains, etc, are examples of GANs. GANs consist of generators and discriminators while generating newer datasets. The role of the generator is to create new datasets, whereas the discriminator attempts to figure out differences from fake images generated by the generator components. The objective of the generator is to develop images that can compete with discriminators without detecting the generative images as fake ones.

2.5 Edge-AI architectures

Edge computing involves edge-level computing nodes, including mobile computing nodes, at different levels of node placement in an edge-AI application. Depending on the underlying characteristics, we could classify the architectures of edge computing applications. The architectures provide an abstract view of involving specific components to accomplish edge-AI tasks. This section details the different edge-AI architectures.

2.5.1 Generic architecture

A generic architecture focuses on multiple application scenarios while designing the software or hardware components of an architecture. This architecture is a naive approach that depicts the core components and mimics the crucial functionalities of edge computing. It is based on a distributed computing environment that focuses on establishing computing continuum based on cloud, fog, and edge nodes.

The important building blocks of a generic architecture of edge computing are listed below:

1. IoT-enabled sensor nodes: Sensor nodes are connected to internet-capable infrastructures either through wired or wireless connections. These sensor nodes may be discovered through manageable networks. Typically, these sensor nodes are low-powered devices and require additional communication modules to communicate with other networks.
2. Connectivity gateways: In order to smoothly communicate with the other networks, sensor nodes are often connected to gateways. Another important challenge with IoT networks is the number of sensor nodes in an IoT-enabled sensor bed. It can be higher in volume and density, so addressing these individual devices can become a challenge. Obviously, these devices need to be addressed with IPv6 addressing protocols. Hence, gateways could map the protocol changes between networks in edge computing architectures.
3. Scalable fog or cloud: Sensor nodes are often inexpensive nodes that cannot undertake huge computations. Hence, a large volume of complex computations are offloaded to scalable computing nodes such as fog nodes or cloud clusters. For instance, a computer vision application requires powerful computing nodes to process video files and images.

Figure 2.4 illustrates the generic architecture of IoT-enabled applications where sensors and actuators are connected to gateways for computing logics using a large array of computing continuum involving edge, fog, and cloud servers.

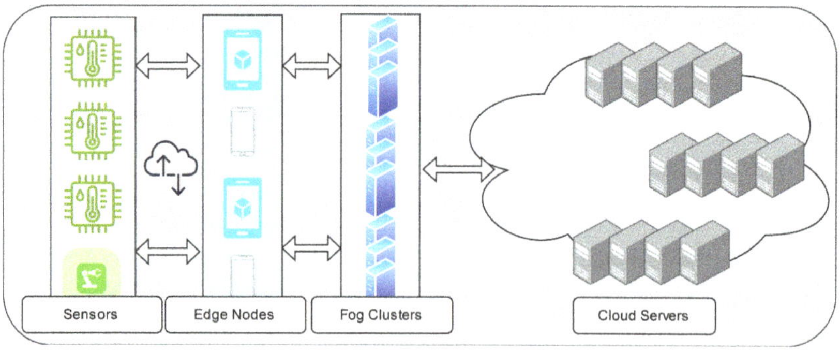

Figure 2.4. Generic IoT architecture.

2.5.2 Layered architecture

Most edge computing architectures are designed in a layered fashion. In some cases, researchers have designed 3-tier to 5-tier architectures to strengthen layer functionalities in applications. It has to be observed that the higher number of layers could increase the computational overhead in applications.

In layered architectures, the placement of compute nodes is organized in the form of layers. For instance, a three-tier layered architecture places sensors or actuators in a layer, gateways in the middle layer, and cloud-based scalable computing environments in the higher layer. A few authors have designed edge-AI frameworks in the form of a things layer, a network layer, a middleware layer, and an application layer [13]. It is also possible to layer each component of edge-AI systems with respect to networking layers such as the physical layer, network layer, transport layer, and application layer.

2.5.3 Service oriented architecture

Edge computing architectures, consisting of distributed computing nodes, can combine several loosely designed services to attain their business goals. These architectures are often centered around data and conceptual logics that are written in the form of services. Programming languages such as Golang, Nodejs, and a few libraries of Python and C++ ripple around newer service-enabled plugins or approaches to satisfy service compositions.

Edge-to-edge, edge-to-fog, edge-to-cloud, and vice versa, the composition of services, symbolizes strong and relentless efforts of service-related developments in such architectures.

Predominantly, service-oriented architectures in edge computing are designed for identifying appropriate learning algorithms, security features, and data sources in real time to execute tasks. This is an important activity of edge nodes as they are limited in space and power sources. For instance, a Raspberry Pi device includes a 32 GB or 64 GB SD card for processing applications. It is not reasonable to store all relevant data required by large applications. Instead, service-level composition of the data or business logics can successfully fetch data or the associated logics at runtime.

It is also crucial to plug in such services to boost the security capabilities of devices, as sensor nodes are often vulnerable to device-level security issues. In addition to security components, the most commonly required services that may be utilized by edge devices are listed as follows:

1. Visualization services: There are hundreds of visualization patterns that could be applied in sensor-oriented applications. For instance, a bar chart, a violin chart, a pie chart, a histogram chart, etc. The majority of applications need to provide data in visualization format to users or clients. It is often difficult to organize all these algorithms or business logics on edge nodes. Visualization services, therefore, may be hosted on nearby compute nodes in the form of services so that they can be utilized by these edge-enabled applications.

2. Debugging services: Remotely debugging services outside of the local environment is complicated due to configurational mismatches and associated issues. In contrast, debugging-related services may be plugged into edge nodes. In this way, a variety of debuggers can be activated multiple times without much memory or computational challenge.
3. Notification services: Sensor-enabled applications are often mounted with many actuators, such as electrical actuators, mechanical actuators, acoustic actuators, etc. Sending emails or warning messages via sensor nodes via edge devices is typically carried out in such IoT-enabled applications. However, modern notification services, or appropriate notification services, vary over time in many applications. In such situations, it is crucial to host different notification services without much burden on the memory requirements of tiny devices.
4. Data processing services: Data collection, data filtration, data interpretation, data storage, and allied tasks can be handled by standalone, separate services. These services are invoked only when required by applications. Hence, tiny devices, when data processing needs to be handled, have to avail themselves of external services hosted on fog or cloud nodes instead of internally storing them.
5. Analytics services: In any IoT-enabled applications, sensors sense environments for measuring properties, and the sensed information is assessed and analyzed by applying learning algorithms, either in distributed or centralized methods. Edge computing prefers such analytics to be carried out on edge devices. In this scenario, the edge nodes have to utilize services that host algorithms as and when required by applications.

Figure 2.5 showcases the list of services that are delivered using edge nodes are shown below:

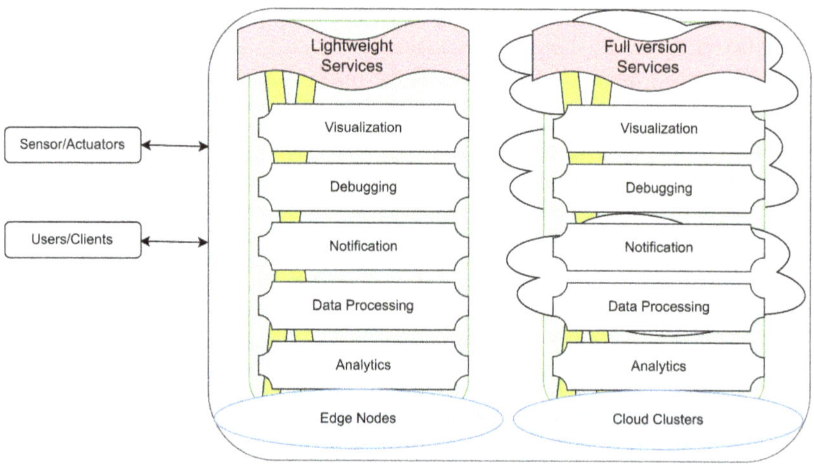

Figure 2.5. Services executed in edge nodes.

2.5.4 Interoperability architecture

Edge computing devices, combined with sensors, are deployed on different platforms supported by multiple vendors. These devices suffer from too many heterogeneity factors, such as policies, communication protocols, device capabilities, etc. There is an urgent need for efficient handling of these diverse characteristics of edge-enabled applications, especially when interoperability is required. The authors of [14] developed a software interoperability architecture that uniquely handles heterogeneous edge devices by a specific messaging pattern.

2.5.4.1 Multi-vendor issues

Devices operated using cellular networks face a long-standing challenge in combining together to frame a single infrastructural network. Edge devices that utilize different networks will have to suffer from reliability issues and hand-off issues when these devices move from one location to another while executing applications. At times, there is a possibility of unintended interruptions that cannot be resumed further in operations. Also, different vendors follow multiple connection procedures and provide hardware support. Consequently, a programmer has to look into these considerations while writing applications for better performance.

2.5.4.2 Communication protocols

There are many communication protocols that are associated with edge-enabled sensor applications. Protocols vary based on supporting networks or infrastructural deployment models. For instance, wireless communications that involve mobile-based cellular networks support protocols such as 4G, LTE, Edge, 5G, 6G, etc. Wireless communications that involve indoor communications support protocols such as Bluetooth, Near Field Communications (NFC), WirelessHART, Zigbee, Indoor LoRa, WiFi, etc. wireless communications that are projected to deal with out-of-door operations excluding cellular networks support LoRAWAN, WiFi, WiFi5, etc. Interoperability of these multi-dimensional communication protocols in edge-enabled computing devices is a challenging aspect for several practitioners.

2.5.4.3 Device capabilities

It is a known fact that the computational and storage capabilities of edge nodes vary. The most notable edge devices, such as the Jetson Nano, Raspberry Pi, ESPs, etc, are manufactured in different configurations. For instance, the majority of edge-capable devices, namely, Jetson Nano and Raspberry Pi devices, have 1 GB, 2 GB, or 4 GB of RAM capabilities. A few devices have high-capacity flash memory units, which differ from the computational service capability of nodes.

In addition to their computational capabilities, these edge-enabled nodes differ in memory requirements and availability conditions.

2.5.5 Metric-oriented edge architecture

Edge computing architectures are illustrated based on performance metrics that are utilized to assess edge-AI applications. Metrics such as throughput, latency, completion ratio, execution time, waiting time, turnaround time, response time, energy consumption, carbon emissions, and costs can tune the characteristics or working behavior of edge AI applications. If an architecture is centered around the performance metrics of applications, it is categorized as a metric-oriented architecture.

A brief note on metrics is given below:

- *Throughput*: This metric expresses the number of tasks completed for a given period of time in an edge-enabled computing infrastructure.
- *Latency*: This metric explains the delay in terms of time that occurs while processing requests in the edge computing infrastructure.
- *Completion ratio*: This metric is useful to assess the capability of hosted computing infrastructure. It describes the total number of requests that are processed by the infrastructure within a specified time period.
- *Execution time*: This metric is the time-related metric that expresses the total time, either in wall-clock time or in CPU-time frame, taken by the services in an edge computing infrastructure.
- *Waiting time*: This is the time a task had to wait before the edge device would be allocated by the infrastructure or platform.
- *Turn around time*: This metric corresponds to the total amount of time a task is spent in the infrastructure to complete the work. This time also includes the waiting time for processing services on different edge nodes in the architecture.
- *Response time*: This metric represents the time when the services are allotted to edge nodes. It is often the case that resources are not allotted due to the limited number of available nodes. And, in most cases, they are busy doing their pre-assigned tasks.
- *Energy consumption:* This metric is utilized to monitor the battery level of edge-associated nodes in an infrastructure. It also relates to the energy drawn from power sources and/or renewable power sources.
- *Carbon emissions*: This metric highlights carbon emissions due to the underlying edge-enabled computing infrastructures. In some architectures, fossil fuels are applied to power the devices, whereas in other architectures, algorithms are designed to diligently offload tasks or services to edge devices that are powered using renewable power sources.
- *Costs involved*: This metric studies the utility of edge devices for an application. For instance, it is better to include Espressif's low-powered ESP than apply NVIDIA's Jetson Nano devices. In some cases, slowing down the edge-related tasks will not hamper the performance of applications, but it can improve the associated costs involved in purchasing such powerful accelerators for the architectures.

Often, the metric-oriented architectures are dynamic in nature, i.e., they tune the applications or edge functions at runtime considering the metrics associated with the frameworks. A detailed explanation of performance issues and the challenges of edge-AI applications is provided in chapter 7.

2.5.6 Open/closed architecture

Edge-enabled architectures are designed based on open source policies and closed policies, depending on the consortium that developed them. Academic organizations jointly formed consortiums to develop open source architectures, which enabled researchers and students to contribute and innovate on newer, modified works. In contrast, closed-source architectures are the type of architecture in which sources are not disclosed to third-party developers. It becomes the proprietary copyright of the company that designs the architecture.

2.5.7 Large/micro/macro-level architecture

Edge-enabled architectures can contain a smaller number of compute nodes or scalable computing nodes to undertake a task. For instance, Amazon offers an AWS Greengrass component to involve as many edge-based devices as possible for applications. The engagement of these components leads to scalable environments due to the combination of edge and cloud infrastructure. Thus, an architecture that includes a larger number of scalable computing environments for improving performance is termed a large-level architecture.

Depending on the scale of computing infrastructure, edge computing infrastructure can be classified into micro- or macro-level architectures. For instance, if one to three edge nodes are utilized in framing an edge-level architecture, then the architecture is termed a micro-level edge architecture. However, if the number of edge devices increases from three to fifteen, the infrastructure is termed a macro-level edge architecture.

2.5.8 Security-specific architecture

In IoT-enabled applications, researchers and practitioners are interested in improving the privacy or security of devices. Most importantly, video frames or face images of edge-enabled applications have to be secured, stored, and processed. An architecture that is designed to protect the privacy of such data using specific algorithms or security mechanisms is considered an example of a security-oriented edge-enabled IoT architecture.

These architectures are often designed with the following design motivations: (a) applying blockchains to prevent data immutability by either machines or intruders [18]. (b) Applying anomaly detection mechanisms to automatically detect security breaches and reporting them to concerned processes or notifiers. (c) Preventing sensors or edge-connected nodes from malware attacks [8] etc.

Figure 2.6 presents the holistic picture of the taxonomy of edge computing architectures that have been implemented by several researchers and practitioners in the past.

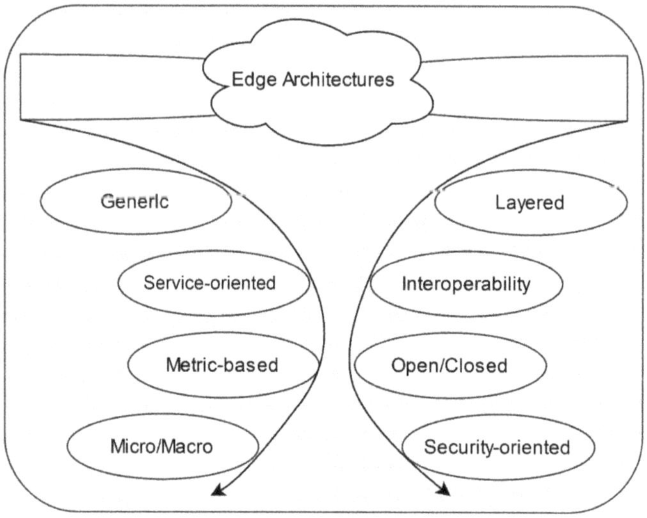

Figure 2.6. Edge-enabled computing architectures.

2.6 Conclusion

This chapter describes the base technologies that are required to learn edge-AI systems. Notably, topics such as IoT, communication protocols including 5G, cloud technologies, and edge architectures. Readers are introduced with these topics so that concepts such as edge-AI frameworks and orchestration platforms are easier to understand.

References

[1] 5 Advantages and Disadvantages of Fog Computing | Limitations & Benefits of Fog Computing—hitechwhizz.com (https://www.hitechwhizz.com/2020/04/5-advantages-and-disadvantages-drawbacks-benefits-of-fog-computing.html) (accessed 27 January 2024)

[2] Ahmad M Z, Adenan A R, Rohmad M S and Yussoff Y M 2023 Performance analysis of secure MQTT communication protocol *2023 19th IEEE International Colloquium on Signal Processing & Its Applications (CSPA)* 225–9 pp

[3] Atmoko R A, Riantini R and Hasin M K 2017 IoT real time data acquisition using MQTT protocol *J. Phys. Conf. Ser.* **853** 012003

[4] Bisht R S, Jain S and Tewari N 2021 Study of wearable IoT devices in 2021: analysis & future prospects *2021 2nd International Conference on Intelligent Engineering and Management (ICIEM)* 577–81

[5] Bormann C, Castellani A P and Shelby Z 2012 CoAP: an application protocol for billions of tiny internet nodes *IEEE Internet Comput.* **16** 62–7

[6] Brownlee J 2024 Networks (GANs)—machinelearningmastery.com (https://machinelearningmastery.com/what-are-generative-adversarial-networks-gans/) (accessed 27 January 2024)

[7] Caiza G, Llamuca E S, Garcia C A, Gallardo-Cardenas F, Lanas D and Garcia M V 2019 Industrial shop-floor integration based on amqp protocol in an IoT environment *2019 IEEE Fourth Ecuador Technical Chapters Meeting (ETCM)* 1–6 pp

[8] Deng X, Tang H, Pei X, Li D and Xue K 2023 MDHE: a malware detection system based on trust hybrid user-edge evaluation in IoT network *IEEE Trans. Inf. Forensics Secur.* **18** 5950–63

[9] Dias I, Ruan L, Ranaweera C and Wong E 2023 From 5G to beyond: Passive optical network and multi-access edge computing integration for latency-sensitive applications *Opt. Fiber Technol.* **75** 103191

[10] Leonardi L, Lo Bello L and Patti G 2023 MRT-LoRa: a multi-hop real-time communication protocol for industrial IoT applications over lora networks *Comput. Commun.* **199** 72–86

[11] Liao Y, Xu Y, Xu H, Yao Z, Wang L and Qiao C 2023 Accelerating federated learning with data and model parallelism in edge computing *IEEE/ACM Trans. Netw.* **32** 904

[12] Machupalli R, Hossain M and Mandal M 2022 Review of ASIC accelerators for deep neural network *Microprocess. Microsyst.* **89** 104441

[13] Minh Q N, Nguyen V-H, Quy V K, Ngoc L A, Chehri A and Jeon G 2022 Edge computing for IoT-enabled smart grid: the future of energy *Energies* **15** 6140

[14] Moreno-Motta J, Moreno-Vera F and Moreno F A 2022 Morarch: a software architecture for interoperability to improve the communication in the edge layer of a smart IoT ecosystem *Smart Trends in Computing and Communications* ed Y-D Zhang, T Senjyu, C So-In and A Joshi (Singapore: Springer) pp 185–95

[15] Spurr R 2022 Smart Manufacturing: Achieve Better Performance with Fanless Edge Devices —community.connection.com (https://community.connection.com/smart-manufacturing-achieve-better-performance-with-fanless-edge-devices/) (accessed 27 January 2024)

[16] Suh K, Kim S, Ahn Y, Kim S, Ju H and Shim B 2022 Deep reinforcement learning-based network slicing for beyond 5G *IEEE Access* **10** 7384–95

[17] Wang H, Xiong D, Wang P and Liu Y 2017 A lightweight XMPP publish/subscribe scheme for resource-constrained IoT devices *IEEE Access* **5** 16393–405

[18] Xue H, Chen D, Zhang N, Dai H-N and Yu K 2023 Integration of blockchain and edge computing in internet of things: a survey *Future Gener. Comput. Syst.* **144** 307–26

[19] Zeng K, Ma Q, Wu J W, Chen Z, Shen T and Yan C 2022 FPGA-based accelerator for object detection: a comprehensive survey *J. Supercomput.* **78** 14096–136

IOP Publishing

Edge Intelligence
Deep learning-enabled edge computing
Shajulin Benedict

Chapter 3

Edge OS and programming models

In general, computers consist of CPU, memory, and I/O units to perform computational tasks. The operating system is software to manage processes that are executed on these compute units. The approach to handling tiny devices such as edge-enabled nodes, including Raspberry Pi, differs both functionally and granularly. This chapter explains the importance of operating systems on IoT-enabled tiny devices; it characterizes the list of available edge OS that are often utilized in various domains involving edge AI solutions.

3.1 Operating systems

Edge nodes, such as Espressif's ESP, Raspberry Pis, etc, have several components similar to computers. Organizing CPU, memory, input units, or output units requires special software, such as operating systems, to increase the performance of edge-enabled applications. The operating systems of edge nodes are based on the hardware modules and running processes. These systems have to diligently serve processes on hardware units, depending on scheduling algorithms and associated performance metrics. This section explains the important hardware components and the features of edge-enabled OS that are popularly utilized for solving edge AI applications.

3.1.1 Hardware components

Operating systems play an essential role in diligently orchestrating hardware components based on processes. They can abstract hardware components to enable virtualization. The hardware components in edge devices can be broadly categorized depending on functional operations such as input, output, or computation.

3.1.1.1 Input units

The devices, such as keyboard, mouse, microphone, joystick, light pen, touch screen, cameras, etc, input data to tiny processing elements in edge nodes. These devices

doi:10.1088/978-0-7503-5593-3ch3

have different data rates and error protection modes that require specialized approaches in operating systems. In general, input devices are capable of transforming user input into machine-readable instructions. The input devices have been modernized in the computing world for decades owing to technological advancements and customer requirements. The role of OS in adopting input devices for providing edge-assisted solutions is the crux of this chapter.

3.1.1.2 Output units

The crucial functionality of output units is to present the data in user-preferred formats. These data emerge as a result of handling instructions in CPU units. The output devices include monitors, headphones, LED/LCD projectors, TVs, printers, 3D printers, plotters, braille embossers, video cards, etc.

The requirements of output devices are as follows: (i) conversion of signals when appropriate to the input of output devices, (ii) smooth transition of data rates between computing machines and the output devices, and (iii) data code verification and error checking mechanisms. The output devices have different data formats and processing capabilities. For instance, analog speakers require analog input sources in a continuous manner; in contrast, digital speakers could not process the continuous data sources as input sources. FreeRTOS or similar embedded OS remain key software to blink LEDs or perform actions.

3.1.1.3 Processing units

CPUs or similar computing units are responsible for processing instructions. In general, the CPU is considered the brain of embedded devices, including edge nodes. It needs to be observed that microcontroller-based edge nodes have limited processing units. The computing elements of microcontrollers include a control logic hardware module, a memory module, and an arithmetic operational module.

The processing units of edge computing nodes are placed nearer to data inputs so that edge-level processing is possible. The CPU-based processing elements fetch instructions from memory units, decode the instructions, execute them using ALU or similar processing units, and store the results for display on the attached output modules.

3.1.2 Functions of edge-based operating systems

As mentioned earlier, an operating system is a software process that interfaces edge devices with IoT-enabled applications. The operating system should be a lightweight implementation with a small amount of code space to accommodate it on embedded devices. It is considered the heart of devices to control and efficiently process the requests of edge AI applications. It has to manage not only hardware resources but also all relevant software processes, including user applications.

In fact, the concept of operating systems has been available for several decades to operate massive computing machines, including powerful servers. The versions of operating systems have evolved over the years in various forms, considering the

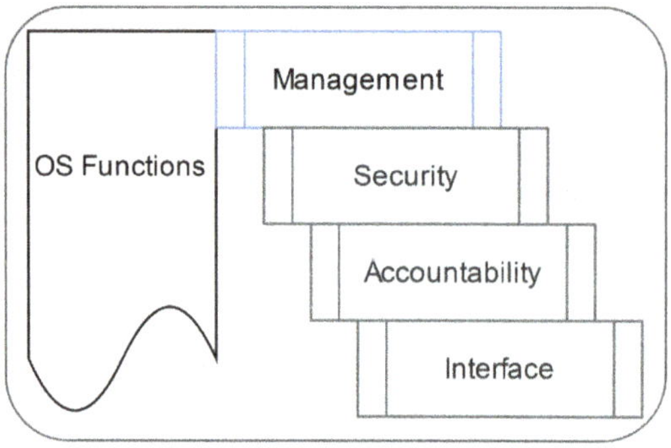

Figure 3.1. OS-related functions in edge devices.

important requirements of working systems. The most important functions (see figure 3.1) of a generic operating system are listed as follows:

1. Management functions: Operating system focuses mainly on managing several important functions of machines. For instance, the management of I/O functions, the management of memory units, the management of CPU processes, the management of files in a file system, etc. The important aspect of the memory management aspect of operating systems is to deliberately allow processes to execute application instructions without delays while considering the space limitations of the memory units. On edge computing nodes, the memory space will be heavily limited, which urges the operating system to provide sophisticated software to handle the associated memory issues. Similarly, handling processes, especially when interrupts occur in edge nodes, is a crucial research aspect that leads to innovations.
2. Security-related functions: Providing security to data on edge systems is one of the most crucial challenges of operating systems. Usually, edge devices are not capable of executing all robust security-related algorithms as they are carried out on computers or clouds. This is because of their lower volume of storage space and computational capabilities. Hence, edge devices are prone to security breaches if they are not handled efficiently. The purpose of operating systems on edge-enabled devices is to ensure lightweight and reasonably acceptable security levels during the execution of applications.
3. Accountability: Auditing the utility of resources is another aspect of operating systems in edge computing environments. In recent years, researchers have proposed the concept of enabling virtualization on edge-enabled devices as similar to traditional servers or data centers. The concept of virtualization is to multiplex hardware resources based on multiple service requests. Obviously, enabling multiple ownership on edge nodes can raise issues relating to utilization patterns and control mechanisms. The edge OS has to have the required features to audit and control the utilization of resources in such multi-tenant environments.

4. Interface: The operating system serves as an interface between users and hardware components such as memory units, I/O, and connected devices. These interfaces are designed in different formats. For instance, some Debian-based operating systems in edge nodes have text-based inputs through terminals, whereas there are a few other input methods in operating systems that utilize click-and-play methods using a mouse or other input devices. In general, interfaces of operating systems have menu items to display running processes, desktop features, and shortcuts that represent different applications. The interfaces must be designed to assist users in a user-friendly way.

3.2 Objectives of OS in edge devices (Edge OS)

The objectives of edge-based operating systems are similar to the normal operating systems of servers. One of the core objectives of designing an edge OS is to abstract hardware resources so that processes are able to access hardware components. Hardware abstraction is important in operating systems to effectively plan resources for queued-up processes; similarly, it increases utilization of resources by selecting an appropriate scheduling algorithm.

The next objective in operating systems is to serve as a bridge between hardware and software processes. In edge-enabled devices, the processes may be interrupted by many I/O modules. These processes have to be scheduled in an agile manner, considering the interrupts. The scheduling algorithms of edge operating systems must be tuned to handle these multiple input devices.

One special requirement of an operating system hosted on edge-enabled devices is to constantly update operating systems with newer features. Application developers keep updating the features based on customer needs. Accordingly, the OS architects have to update the OS. There are updates and version changes in the newer releases of operating systems. For instance, FreeRTOS has features to automatically update the software features of devices using over-the-air (OTA) updates.

Another objective is to handle requests at local devices without interrupting globally connected nodes. In this way, the resources could be optimized while securing the privacy levels of applications. This feature is quite beneficial for several large-scale IoT-enabled applications, including digital forensics.

Also, while designing operating systems, including edge-enabled OS, developers attempt to increase the utility rate of the outcome. For instance, an operating system should have a module-programmable structure so that the framework will be convenient for future researchers or developers in the future. The software solution must have sufficient convenience for not only the developers but also the users. Notably, the users should feel comfortable selecting configurations or enabling or disabling hardware components. In [14], the authors have listed the crucial characteristics of operating systems that need to be hosted on edge-enabled devices.

3.3 Taxonomy of OS: toward edge

Traditionally, OS is classified into batch OS, time-sharing OS, distributed OS, real-time OS, embedded OS, and networked OS. Batch OS is meant to process tasks in batches. The processes are not instantly scheduled with resources. Instead, the selective processes are collected from a list of users and scheduled based on the availability of resources at a particular instant in time. In batch OS, the OS does not have a direct connection to resources. Hence, the time required to complete the process is unknown or unpredictable. However, the batch OS approach is efficient in handling a large number of users with limited resources.

In a time-sharing OS, the tasks are processed in a time-sharing manner [10]. This offers a fair scheduling of tasks in a computing system, as users are allotted a specific time frame for all executions. It is an interactive system when compared to the previous batch OS system. This means users are allowed to execute tasks without interference from other users during the specified time frame.

The key to developing a distributed OS [17] is to ensure transparency of resources as the resources are spread across regions or locations. Some resources are connected through wide-area networks spanning multiple administrative domains. Handling the task requirements of users by providing resources across regions with the utmost transparency is crucial in a distributed OS system. Obviously, there is a possibility of lagging in resource availability or resource distribution, or there is a requirement of replicating resources considering the performance aspects of applications. Additionally, it has been observed in the past that resources were shared by different vendors. Hence, the sharing of resources has to follow open standards to ensure fault tolerance.

Real-time OS [6] mainly focuses on real-time operations where the input tasks vary over time. The input tasks will require hard deadlines, such that the real-time OS subjects will adjust the schedules depending on the changing demands. The OS is designed to handle multiple simultaneous requests from users when compared to the time-sharing OS or batch OS mechanisms. In principle, the real-time OS requires only a limited amount of storage space, as most of the requests are processed immediately depending on the availability of time and resources. In most cases, the real-time OS is designed for a specific application or process. In contrast, the majority of the other available OSs are designed to handle general-purpose applications or processes.

Embedded OS [2] is designed for embedded devices that are customized for specific applications. As the name indicates, the OS is loaded into one chip, preferably microcontrollers. The embedded OS undertakes real-time operations or processes similar to the real-time OS. However, it is not so strict in time-bound applications. The embedded OS mainly focuses on things or objects, such as IoT-enabled nodes, rather than computers for processing lightweight processes. It has to deal with lots of I/O interrupts. For instance, Espressif's ESP 32 may have to handle 26 general-purpose I/O units or sensors using its associated operating system. This points out the importance of configurational support and interrupt handling procedures in such operating systems when compared to the majority of other general-purpose operating systems.

Apart from these types of OS, there is also a networked OS that specializes in network-related devices such as switches, routers, or gateways. Since these types of operating systems are designed for specific devices, they are not appropriate for general-purpose computers. The key features of a networked OS include support for multiple communication protocols and standards. For instance, a device supporting the TCP/IP standard has to communicate efficiently with Zigbee-based communication protocol devices using networked OS systems. Also, networked OS support enables the sharing of common file systems and databases with ease when compared to other operating systems.

Edge OS is a specialized operating system that falls under the categories of real-time OS and embedded OS. The operating system of edge computing nodes is meant to process interrupts and deliver results in real-time. The OS must have default lightweight security options so that connected devices or things will be in a position to protect them from external security challenges.

3.4 Edge OS: examples

Operating systems in edge computing nodes are classified depending on their capabilities. A server-based edge node can host general-purpose operating systems, whereas a tiny, power-constrained edge device must have an embedded or real-time OS. Also, based on hosting the operating systems on networked devices such as routers, the operating system is named a networked OS. In this section, we shall understand the most commonly available operating systems for edge computing nodes, along with their original features.

Figure 3.2 highlights the top 12 operating systems that are utilized in edge-AI systems. They are detailed in the following subsections.

3.4.1 QNX neutrino OS

QNX is pronounced quenix. It is a UNIX-based operating system. However, it is not exactly the UNIX product. It is a micro-kernel version of the operating system designed for Blueberry devices. It offers synchronous inter-process communication using a message-passing approach [3]. Initially, the operating system was designed by a Canadian company as a real-time operating solution for tiny devices. Later, the neutrino micro-kernel version of the operating system was designed by the Blackberry company to provide the required features of a tiny edge device.

The most important features of QNX Neutrino OS are listed as follows: (a) real-time: QNX Neutrino OS is designed to support real-time operations with ease. This enables QNX neutrino to quickly act upon the I/O interrupts of connected devices. (b) Monitoring: QNX neutrino has a self-monitoring feature in the operating system. The nature of this feature is to automatically monitor the status of processes or threads so that a few recovery mechanisms are activated through lightweight kernels. (c) Scheduling: QNX Neutrino operating system supports multiple scheduling options, such as symmetric or asymmetric scheduling options for running

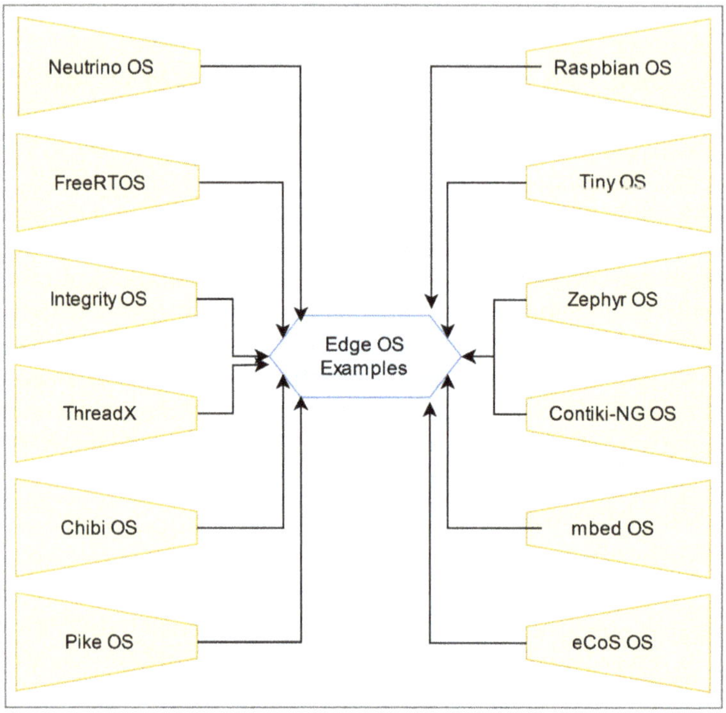

Figure 3.2. Edge-OS: a1 few examples.

processes. In fact, scheduling is one of the key features of operating systems such as QNX Neutrino. For instance, the authors of [5] have studied the efficiency of applying an adaptive partitioning scheduling algorithm to minimize the latency of autonomous applications. (d) Security: the operating system is designed to inherently support security challenges. Notably, security issues such as memory leaks, process hacking, file system failures, etc, could be minimized based on the features enabled in the operating system. In QNX neutrino OS, the security components are implemented in a layered approach that includes system-level kernel modules for mitigating security challenges.

3.4.2 FreeRTOS

FreeRTOS is a real-time operating system designed for embedded systems [1]. The OS is developed based on the C programming language with minimal storage space requirements. The OS is installed on several microcontroller boards of IoT-enabled devices, such as the Arudino UNO, Raspberry Pi Pico, ESP 32, Jetson Nano, etc.

Currently, FreeRTOS is widely utilized in the IoT market due to its simplicity and modular implementation capabilities. It has wide licensing options, such that commercial or academic products are based on FreeRTOS. In order to attract a wide sector of researchers or practitioners, a large ecosystem based on FreeRTOS has been developed in the recent past. For instance, FreeRTOS has been extended to offer a trace of process utility using the 'trace' component; a TCP and an UDP component have been developed based on FreeRTOS to support reliable continuous transmission and unreliable communications. Similarly, FreeRTOS has enabled the development of tens of thousands of IoT-based products [18].

3.4.3 Integrity OS

Integrity OS is an operating system [7] developed by GreenHills Software, Inc. The OS is designed as a real-time operating system with more emphasis on reliability and security. The OS is designed to offer a safe real-time operation without compromising the performance of applications. It supports several file systems, such as FAT, NFS, and journaling features of file systems. With the inherent routing features, Integrity OS is able to support long-range communication-based web services such as HTTP, sHTTP, SOAP, AJAX, JSON, etc.

An important feature of the Integrity OS is to enhance reliable operations for mission-critical applications. For instance, a military application can continue its operation as long as the OS offers sufficient CPU or memory resources even when the resources are not available. This means that a special reservation is kept in mind while delivering the resources to running processes. Notably, to prevent memory stack overflow in mission-critical applications, the Integrity OS has a dedicated memory provisioning unit at the kernel level. In doing so, the applications could continue their operations without failure due to the non-availability of resources.

3.4.4 ThreadX

ThreadX is a product of Microsoft [15]. It was developed to provide an embedded operating system for cloud-based IoT-enabled applications. The operating system is designed to satisfy multiple embedded architectures that fall under RISC, CISC, or DSP processors. The peculiar feature of the ThreadX OS is the picokernel feature, which brings together multiple embedded OS functions in the form of services. Compared to several other embedded OS, this architecture has increased utility among embedded researchers owing to its serviceable approach. For instance, the core OS features will be available to layer together multiple other service-oriented functional components as needed.

The most important features of ThreadX OS are: (i) to schedule processes in a pre-emptive or cooperative manner, (ii) to handle interrupts efficiently at run-time, (iii) to iteratively process event-based requests for tasks, etc.

To manage Boolean operations, event flags, mutex actions, threads, semaphores, block operations, and other tasks, the ThreadX operating system offers a variety of data types. Specialized APIs created for this reason handle the intended functions of operating systems. To halt and release a block pointer, for example, use the following:

```
/* To create mutexes… */
tx_block_release(VOID *block_ptr);
```

```
tx_mutex_create(TX_MUTEX *mutex_ptr,
CHAR *name_ptr, UINT inherit);.
```

In the ThreadX operating system, mutex activities can be stopped or deleted by API calls, just like they can be created.

In ThreadX, device drivers have been created to support devices from various vendors. Since ThreadX supports these device drivers as services rather than stacked modules, initializing them can be difficult. In ThreadX, specific driver components are designed to manage device input and output data. Buffering is turned on in certain ThreadX operating system driver-related services in order to do this.

3.4.5 PikeOS

PikeOS is a real-time operating system designed by the Sysgo Embedding Innovations company [8]. It supports architectures such as x86, POWER, ARM, and RISC-V. An important feature of PikeOS is the virtual provisioning of resources in real-time. Compared to many other existing real-time operating systems, PikeOS specializes in supporting virtual instances of resources using hypervisors in varied architectures. Due to the involvement of virtual instances using hypervisors, PikeOS enables the migration of resources or services across multiple hardware units. At the time of writing this book, PikeOS 5.1.3 was developed to assist system-on-chip devices. Being a product of the European Union, the approach is open to all practitioners without any geographical restrictions.

3.4.6 Chibi OS

ChibiOS is a real-time operating system developed by the ChibiOS group as a project. The implementation of ChibiOS is focused on Linux-based machines under open-source licenses. Obviously, it enables researchers to experiment with newer scheduling policies or algorithms and learn interrupt handling mechanisms on embedded devices. Additionally, people have explored the importance of stacks and stack memory sizes while experimenting with embedded applications.

The internal states of ChibiOS are classified into seven major states [4]: init state, normal state, suspended state, disabled state, sleep state, locked state, serving state, and halted state. The init state is the beginning state of operations. It is literally referred to as a reset operation. The normal state is the state when all APIs are accessible by resources. The suspended state occurs when the API calls become inaccessible to perform operations. During this state, the processes are not allowed to perform any actions. The serving state is equivalent to the running state of any operating system. During this state, the APIs are accessible for execution. When compared to the normal state, the serving state is involved in utilizing the resources. The halted and locked states are meant to stop the running processes. In the halt state, the services are stopped using specific API calls, whereas the locked state doesn't allow any processes to override the current running situation. Notably, it is involved when the files need to be protected from other external modifications.

3.4.7 Raspbian OS

Raspbian OS is an operating system designed for Raspberry Pi devices. It has over 35 000 packages that are pre-compiled for many devices. Accordingly, there are several use-cases or applications based on Raspberry Pi devices [13]. The versions of Raspbian OS are designed for powerful computational units, or SoC devices. In the latter case, the operating system does not involve all features of desktop display functions. In contrast, a lite version of most of the functions of operating systems is available on SoC devices. For instance, Raspberry Pi Pico devices include the lite version of the Raspbian OS.

In general, the Raspbian OS is based on the Debian-based Linux operating system. It is mainly designed for computationally powerful devices rather than many other embedded operating systems that were discussed in the previous sections. The intention of utilizing Raspberry Pi boards is to handle multiple sensors or actuators, including motors, using the assistance of the Raspberry OS. The OS is loaded into the microSDcards that have 16 GB to 128 GB of flash memory.

3.4.8 TinyOS

TinyOS is an embedded OS designed by Stanford University as a BSD license version under the open-source category [12]. The TinyOS has a collection of drivers and schedulers for connecting sensors or actuators to microcontrollers. The space complexity of TinyOS is minimal when compared to many other embedded operating systems. However, the safety features incorporated in this OS are much less than in dedicated operating systems such as Integrity OS. This OS is designed for several home-specific embedded applications that involve fridges, washing machines, heaters, air conditioners, etc. In addition, they are suitable for several smart devices in buildings, industries, or manufacturing units. TinyOS is written using the nesC language, a dialect language based on the C programming language.

The most important features of TinyOS include setting timers, scheduling tasks in real-time, abstracting resources, enabling virtualized resources, managing power utilization, handling I/O devices, etc. At the networking level, the OS has support to handle multi-protocols such as WIFI, Bluetooth, serial communications, IEEE 802.15.4 protocol stacks, etc. Specific recommendations have been floated to support the standardization of protocols; for example, TEP125 is formulated to standardize the frame structures of inter-node communications among microcontrollers that utilize TinyOS operating systems.

3.4.9 Zephyr OS

Zephyr OS is developed based on the Apache license, which has a minimal code space. Typically, this OS is designed for IoT-enabled tiny devices where storage space limitations are a particular concern. The OS extends its support to several architectural devices belonging to ARM, Intelx86, Sparc, Tensilica Xtensa, RISC-V, Sparc, etc. The OS offers services in broader categories. For instance, it offers kernel-level services to support multi-threading, power management, memory allocation, synchronization, and interrupt handling services; it provides several scheduling-algorithmic services such as cooperative scheduling, earliest deadline first scheduling, time-sharing, and queue-based scheduling; it enables memory protection by preventing stack overflows, permission violations, etc; it extends support to varieties of device drivers to drive devices belonging to different architectures; it delivers a strong isolation feature when deployed on embedded nodes [9], etc.

Zephyr OS is based on the C programming language and the C-Make build system. It engages the developers to constantly innovate by extending the tool or combining the OS with novel solutions. For instance, visualization tools such as Trace Compass, I/O Taxonomy, Object Tracking, etc, are offered with the specialized features of the Zephyr OS. In addition, it has modernized support for Bluetooth 5.0 to enable supportive features to handle low-energy modules.

To minimize the power utilization of devices, Zephyr OS has inbuilt power management mechanisms, i.e., the kernel enters an idle state while resources are not utilized due to the non-availability of data, etc. The power management policies or allied control subsystems are powerful enough to decide whether to switch off the devices during the run-time of applications.

3.4.10 Contiki-NG OS

Contiki-NG OS is designed specifically for memory-constrained IoT-enabled applications and corresponding memory-limited tiny devices. It is an open-source operating system that can invite multiple developers to demonstrate IoT-enabled products [11]. It mainly focuses on minimizing the energy utilization of individual components of devices through diligent energy-efficient architectures. It has support for several energy-conscious architectures belonging to different categories, such as ARM Cortex and Texas Instruments. As Contiki-NG OS places its emphasis on tiny IoT-enabled devices, the code space is limited to 100 K.

Contiki-NG supports networks by offering low-power IPv6 addressing and communications. Due to the utilization of IPv6, it has wider support for IP-based devices when compared to several other incompatible protocols, such as Zigbee. The core features of Contiki-NG OS involve offering real-time timers, multi-tasking, scheduling, packet buffering, energy monitoring, etc. Assistance to reliable communications is limited in Contiki-NG OS-based devices owing to UDP-based communications.

In order to monitor the energy consumption of the compute modules of Contiki-NG, the OS utilizes the Energest tool. This tool is a software module that keeps track of resource utilization in real-time. Depending on the poor utilization of devices, the Energest tool could turn off a hardware module. There are several APIs associated with the OS that automatically control the energy efficiency of memory-constrained IoT nodes.

3.4.11 mbed OS

mbed OS is an operating system designed for ARM-based cortex-M devices. In general, ARM devices are utilized for edge computing applications. The mbed OS is a specialized OS for IoT-enabled connected systems. This is due to the fact that it offers sufficient security features and modular development environments. It is a C/C++-based modular architecture that is specifically designed for resource-constrained devices.

The recent version v6.15 has the feature that it is able to compile sources online and develop code using online Mbed Studio. The Mbed Studio is an integrated development environment that checks the program's security features, including memory leaks. It has the capability to migrate virtual resources across different edge computing nodes by abstracting resources through hypervisors. To do so, mbed OS offers a specialized set of migration tools using the mbed-tools of MBED-CLI2.

The support for Windows and MacOS machines, in addition to the majority of Linux-based installers, has increased the popularity of utilizing Mbed OS in many IoT-enabled applications. The development platform has additional support for the majority of the existing integrated development tools, such as Eclipse, Visual Studio Code, etc.

BLE nodes are supported with a wide variety of GATT implementations so that IoT-enabled resource-constrained nodes could be efficiently handled using this operating system. For instance, body area network-based nodes can be connected by advertising the availability of nodes. Similarly, the OS has a wide applicability in communicating with many communication protocols due to the availability of assistive programs [16].

3.4.12 eCoS OS

eCoS operating systems are designed to consider the embedded nature of devices. It is a configurable software resource that has only one process at a time. However, multiple thread-level lightweight processes are provided in this operating system to increase the performance of real-time embedded applications.

The operating system is written in C++. In addition, it has the capability to support several C-based software solutions. It has real-time support for handling interrupt-based tasks. Due to its free open-source versions, eCoS OS is widely experimented with in developing a novel OS and its features.

3.5 Process states in edge-OS

A process, in general, is a program in execution. A program is loaded into memory and executed by providing state information so that we can manage these processes. In general, a traditional OS maintains several states for managing processes. However, embedded OS-based devices such as the ESP 32 have limited process states due to the limited computational capabilities. In general, the process states of embedded devices are classified into four broad categories. They are listed as follows:

1. Ready state: In this state, resources are created for the execution of the processes. The resources include threads, memory units, and other resources. The ready state is the indication for entering the running state of the device in which the CPUs are allotted for execution. This state is governed by the

scheduler of the operating system. In some cases, the state of the program is switched between running processes.

2. Running state: A program in execution is termed a running state. The instructions have to be executed using specific CPU, memory, or computing units. During the running state of a process, the execution of the process has to be carried out intermittently unless any interrupts or hardware failures occur.
3. Blocked state: A process undergoes a blocked state if it waits for data to be processed or instructions to be executed. A process can be purposefully delayed by programmers due to the anticipated non-availability of resources or input data. In such situations, it is important to set the processes to a blocked state so that the other processes can utilize the resources. The advantage of the blocked state of a process is the capability to resume operations after a period of time. For example, the delay(2) function of a process blocks the process for 2 seconds and resumes on its own.
4. Suspended state: A process is considered to be in a suspended state when the instructions are not executed for a specific time period. It is similar to the blocked state of the processes. However, the suspended processes could not resume on their own, i.e., there is no specific time period to stop and start the processes. But it waits for another process to resume its operations.

3.6 FreeRTOS: OS for embedded devices

FreeRTOS is one of the popular operating systems that are utilized in embedded-based devices. FreeRTOS is a lightweight operating system that manages processes, schedules tasks, and handles interrupts. In this section, we learn about the application of FreeRTOS to esp32-based processors.

3.6.1 FreeRTOS: ESP 32

FreeRTOS is easily ported to esp32-based microcontrollers due to its support for integrated development environments such as ESP-IDF. ESP 32 microcontrollers, such as ESP 32-WROOM boards, have two cores that can handle multiple tasks simultaneously. In order to support multi-tasking on smaller embedded boards, ESP-IDF has integrated FreeRTOS natively into the IDE. The notion is to ease users, especially programmers, into utilizing operating systems instead of directly programming devices.

The advantages of including FreeRTOS in ESP 32-based boards are listed as follows: (a) integration of devices and applications, (b) executing parallel tasks, (c) efficiently utilizing timers, (d) organizing semaphores, (e) dealing with multiple communication protocols such as WiFi, Bluetooth, BLE, HART, etc, (f) initiating queues of tasks, (g) scheduling tasks using sophisticated algorithms, etc.

FreeRTOS and associated functions that create tasks or schedule functions are implemented in the form of APIs so that users could invoke the functions in their functional logics. Initially, users have to import appropriate libraries defined in header files, such as:

```
#include <freertos/FreeRTOS.h> // for FreeRTOS
#include <freertos/task.h> // for tasks
```

3.6.2 Parallel tasks

Due to the availability of two cores in ESP 32-S boards, multiple tasks are executed in parallel in such microcontrollers. In this subsection, we explore the benefit of executing FreeRTOS-based tasks for a subset of Espressif boards.

3.6.2.1 Task creation

Tasks are defined and created using function defined below:

```
 xTaskCreate(Task1_ReadSensor, "Reading Sensor", 4096,
NULL, 10,
 &TaskHandle1);
 //arg-1 -> name of the task
 //arg-2 -> description of the task
 //arg-3 -> stack size
 //arg-4 -> parameters that need to passed to the task
 //arg-5 -> priority of the task
 //arg-6 -> status of the task (e.g., created, deleted,
suspended,
 resumed, …)
```

The `xTaskCreate` function specified above creates a task to read sensor values that could be defined in a user-specific function. It has arguments to specify the name of the task, description, stack size, parameters of tasks, priority, and the current status. The parameters of tasks could be passed on while executing them by cores. The status of the task is defined as either created, deleted, suspended, resumed, or blocked; it is stored in a user-defined pointer named `TaskHandle1`.

Creating task is invoked within app_main() function of embedded applications, as given below:

```
void app_main(void)
{
xTaskCreate(Task1_ReadSensor, "Reading Sensor",
4096, NULL, 10, &TaskHandle1);
}
```

3.6.2.2 Pinned tasks

Tasks of applications could be pinned to a specific core: i.e., core 0 or core 1: using specialized APIs. Notably, the following function performs actions that are defined in a function named `Task2_EnableActuator` in core 1. All instructions of the function are pinned to core 1 if they are mentioned using the `xTaskCreatePinnedToCore` function.

```
            xTaskCreatePinnedToCore(Task2_EnableActuator,
 "Performing Action",
  4096, NULL,10, &TaskHandle2, 1);
  //arg-7 ->pinned to CPU core 1
```

It could be observed from the following program that task 2 will be executed on core 1 all the time, irrespective of the availability of core 0.

```
#include <stdio.h>
#include <freertos/FreeRTOS.h>
#include <freertos/task.h>

TaskHandle_t TaskHandle1 = NULL;
TaskHandle_t TaskHandle2 = NULL;

void Task1_ReadSensor(void *arg)
{
while(1){
/*… Write code to read sensor values …*/
printf("Task 1 is at %d… \n", xPortGetCoreID());
vTaskDelay(1000/ portTICK_PERIOD_MS);
}
}
void Task2_EnableActuator(void *arg)
{
while(1){
/*… Write code to perform actions …*/
printf("Task 2 is at %d… \n", xPortGetCoreID());
vTaskDelay(1000/ portTICK_PERIOD_MS);
}
}
void app_main(void)
{
xTaskCreate(Task1_ReadSensor, "Reading Sensor",
4096, NULL, 10, &TaskHandle1);
          xTaskCreatePinnedToCore(Task2_EnableActuator,
 "Performing Action",
4096, NULL,10, &TaskHandle2, 1);
}
```

3.6.2.3 Task suspend/resume

Tasks are suspended or resumed based on functions given below:

```
vTaskSuspend(TaskHandle2); //Task 2 is suspended
vTaskResume(TaskHandle2); //Task 2 is resumed
```

FreeRTOS could identify suspended tasks or resumed tasks using handles. Programmatically, the tasks could be created, deleted, resumed, or suspended due to specific functions that are developed in the FreeRTOS operating systems. In doing so, the programmers are capable of effectively handling parallel tasks while considering the events that evolve from the connected sensors/actuators. Similar to these functions, edge AI developers can utilize the functions of operating systems while designing their applications.

3.7 Conclusion

Operating systems play a crucial role in embedded devices. This chapter explores a few potential operating systems that are suitable for enabling edge-AI applications. In addition, it showcases a few examples, such as creating parallel tasks and pinning them to specific cores of embedded nodes. Readers can learn the application of operating systems by practically working with open-source operating systems such as FreeRTOS.

References

[1] FreeRTOS–Market leading RTOS (Real Time Operating System) for embedded systems with Internet of Things extensions–freertos.org (https://www.freertos.org/) (accessed 27 January 2024)

[2] Afonso F, Silva C, Montenegro S and Tavares A 2007 Applying aspects to a real-time embedded operating system *Proceedings of the 6th Workshop on Aspects, Components, and Patterns for Infrastructure Software ACP4IS '07* (New York, NY: Association for Computing Machinery) pp 1–es

[3] Becker M, Dasari D and Casini D 2023 On the QNX IPC: assessing predictability for local and distributed real-time systems *2023 IEEE 29th Real-Time and Embedded Technology and Applications Symposium (RTAS)* pp 289–302

[4] ChibiOS 2022 ChibiOS: Kernel Concepts—chibiforge.org (https://chibiforge.org/doc/21.11/full_rm/concepts.html#system_states) (accessed 27 January 2024)

[5] Dasari D, Becker M, Casini D and Blass T 2022 End-to-end analysis of event chains under the qnx adaptive partitioning scheduler *2022 IEEE 28th Real-Time and Embedded Technology and Applications Symposium (RTAS)* pp 214–27

[6] Hambarde P, Varma R and Jha S 2014 The survey of real time operating system *2014 International Conference on Electronic Systems, Signal Processing and Computing Technologies* pp 34–9

[7] GHS Inc. 2024 INTEGRITY Real-time Operating System—ghs.com (https://www.ghs.com/products/rtos/integrity.html) (accessed 27 January 2024)

[8] Sysgo Embedding Innovations 2013 PikeOS RTOS & Hypervisor | SYSGO—sysgo.com (https://www.sysgo.com/pikeos) (accessed 27 January 2024)

[9] Li Y and Takada H 2022 iSotEE: a hypervisor middleware for IoT-enabled resource-constrained reliable systems *IEEE Access* **10** 8566–76

[10] Oestreicher M D, Bailey M J and Strauss J I 1967 George 3–a general purpose time sharing and operating system *Commun. ACM* **10** 685–93

[11] Oikonomou G, Duquennoy S, Elsts A, Eriksson J, Tanaka Y and Tsiftes N 2022 The Contiki-NG open source operating system for next generation IoT devices *SoftwareX* **18** 101089

[12] TinyOS 2012 TinyOS Home Page—tinyos.net (http://www.tinyos.net/) (accessed 27 January 2024)

[13] parab J 2023 Simple applications with Raspberry Pi Python *Programming Recipes for IoT Applications* (Singapore: Springer) pp 49–87

[14] Patel B and Shah P 2022 Operating system support, protocol stack with key concerns and testbed facilities for IoT: a case study perspective *J. King Saud Univ. Comput. Inf. Sci.* **34** 5420–34

[15] Azure RTOS 2024 Chapter 1—Introduction to Azure RTOS ThreadX—learn.microsoft.com (https://learn.microsoft.com/en-us/azure/rtos/threadx/chapter1) (accessed 27 January 2024)

[16] Strelkov N O, Krutskikh V V and Shalimova E V 2022 Programming STM32 nucleo platform for IoT education using STM32duino and Mbed OS *2022 VI International Conference on Information Technologies in Engineering Education (Inforino)* pp 1–6

[17] Tanenbaum A S and Van Renesse R 1985 Distributed operating systems *ACM Comput. Surv.* **17** 419–70

[18] Wang B and Seltzer M 2022 Tinkertoy: build your own operating systems for IoT devices *IEEE Trans. Comput.-Aided Des. Integr. Circuits Syst.* **41** 4028–39

Part II

Learning techniques

Chapter 4

Edge intelligence: learning techniques

Edge intelligence has been promulgated in multiple areas by several researchers and practitioners in the recent past. This chapter explores the most commonly applied techniques in order to motivate readers to develop newer learning methods or federated procedures. The entire chapter is dedicated to explaining different edge-assisted learning techniques that are applied while practicing IoT-specific applications.

4.1 Edge intelligence: a need

Learning is predominantly practiced in almost all activities. In general, a human brain consists of over 100 billion neurons to capture information from various sensing sources. The three major parts of the brain, the cerebrum, cerebellum, and stem, play a key role in processing the captured information from different sensory modes, including touch sensory information. Notably, each portion of the brain has specific functionalities to process learning tasks, i.e., visual learning and inference happen at the occipital lobe of the brain [4].

The sensory information is transmitted through the sensory organs of the body by synapses along the neural paths. They are stored in the short-term memory portion of the brain. However, depending on the frequency of access to that information, the content is transferred to the long-term memory unit of the brain. Long-term memory involves the structural core of the brain.

In human brains, memory formation is based on connections between neurons. The connections are either strengthened or weakened, depending on several factors. For instance, information associated with emotions or repetitions can establish strong connections between neurons. Similarly, information connected with multiple sensory inputs such as touch, smell, sound, sight, and taste can lead to a strong neuron connection.

Instead of processing the information in the brain, there is the possibility of performing reasoning or decision-making actions nearer to the sensory parts of the

doi:10.1088/978-0-7503-5593-3ch4

body. For instance, touching a hot object suddenly makes us move away if it is painful. In this case, the sensory information is processed nearer to the spinal cord rather than sending all the information to the brain before an action is performed. The core concept in this scenario is the involvement of edge-level intelligence in humans.

Similar to human brains, learning processes could be carried out by edge-level machines to deliver prompt learning inferences. Notably, performing edge-level intelligence is heavily practiced while executing deep learning algorithms on compute nodes that reside nearer to sensor nodes. In doing so, we are able to reduce the network latency of IoT-enabled applications and deliver actions much more quickly than traditional cloud-centric IoT solutions.

The application of learning technologies using edge-level nodes is carried out in various approaches as shown below:

1. federated learning (FL);
2. DNN split learning;
3. transfer learning;
4. gossip learning.

The core concepts of these edge-inclined learning methods, their pros and cons, are discussed in detail in the following sections.

4.2 Federated learning

Traditional machine learning algorithms and practices have limited extensibility due to the large volume of data from various sources while modeling data. Having seen remarkable growth in the AI industry, researchers have come up with innovative solutions that combine multiple data sources from various geographical locations to develop learning models. In FL, a learning model is developed collaboratively using edge-level compute nodes.

Unfortunately, combining data sources from different locations is prone to privacy violations [11]. It has increased the need to develop FL concepts. In FL, the models are distributed across regions; the models are developed based on the data sources that are closer to the privacy-related locations. Later, they are transferred to centralized clouds.

For instance, a word processing application could predict the preferred words in a sentence tailored to the users. To do so, users utilize edge-enabled devices to develop models and transfer the extracted features in an encrypted fashion to clouds. The clouds collect model snippets from different end users to develop a more robust learning model with refined predictions. Later, the edge-level models could be updated using the updated learning models of clouds. In this way, the predictions or classifications are based on a joint learning model developed in combination with data nearer to the users and generalized observations from various locations. The authors of [7] have studied the applicability of transforming learning models from traditional machine learning to distributed systems.

4.2.1 Federated learning architecture

In general, FL architecture involves multiple edge-level nodes to perform learning based on regional data sources. The observed features are transferred to clouds to generalize the findings in a consolidated manner.

Figure 4.1 expresses a generic FL architecture. Here, it can be observed that a global learning model is created by the involvement of several compute nodes, including edge-level nodes. The local data from edge locations is utilized to create local models. Later, they are shared with the cloud-based central cloud clusters to generate an FL model. This model has aggregated information from several local compute nodes.

4.2.2 Federated learning approaches

Broadly, we can classify the FL mechanism into the three broad approaches discussed in the following sections.

4.2.2.1 Data/model-transfer approach

In most cases, data or models are transferred between the edge-enabled learning device and the centralized cloud-based servers. Depending on the transferred information, the FL varies. For instance, if the modeling parameters are transferred across devices irrespective of data, it is horizontal FL. In this case, independent training is carried out with the regional dataset. Hence, the modeling parameters are sufficient to be transferred to the decentralized edge devices; if the training dataset remains the same for distributed edge learning nodes irrespective of modeling parameters, it is vertical FL. In this case, the training dataset remains the same in all regions. However, the prediction components utilized in different locations might

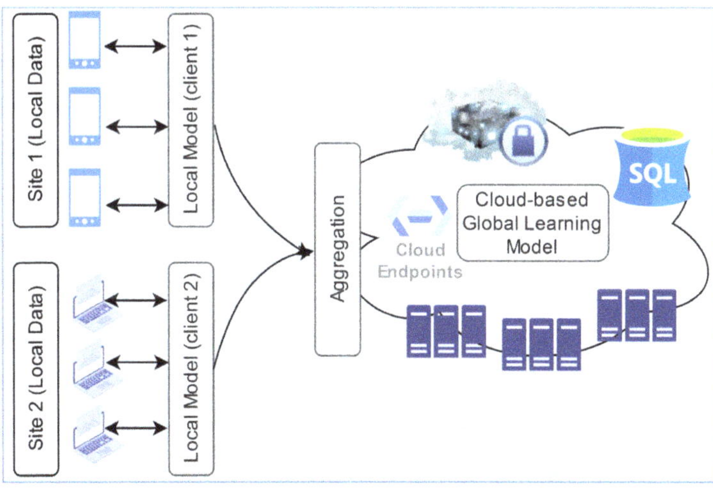

Figure 4.1. FL: a generic architecture.

vary depending on the requirements, and if models are transferred across edge devices to perform the learning process, it is model-transfer FL.

4.2.2.2 Meta-information transfer approach

In specific cases, it is not required to transfer the models or dataset across edge nodes. It might be too expensive to transfer them. In the meta-information transfer approach, the FL mechanism transfers only a few meta-datasets or models between the decentralized edge nodes. In doing so, the communication latency between devices is minimized. However, this FL approach does not fit well with a few subsets of applications or use cases.

The meta-information transfer approach requires the underlying networks to be hierarchical in nature, i.e., levels of connected computing network architecture are necessary to process the meta-information.

4.2.2.3 Reward-specific federated learning approach

Rewarding learning agents while developing learning models is another specific variety of FL. In the reward-specific FL approach, rewards are provided to algorithms when they perform well for different datasets. In general, the accuracy of learning algorithms, especially in a decentralized FL environment, depends on the data utilized in federated locations. This means that the rewards can vary depending on the data inputs to the FL system.

4.2.3 Federated learning challenges

Although FL is quite beneficial for developing a decentralized learning model using edge nodes, it faces a few challenges. Figure 4.2 depicts these challenges identified by researchers in the past.

The details about these identified challenges are described below:

1. Model convergence: The convergence of learning models is a challenging aspect of FL. This is due to the fact that the models have to express the findings or share the extracted features between decentralized computational devices. The format mismatch or configurational issues of models can lead to

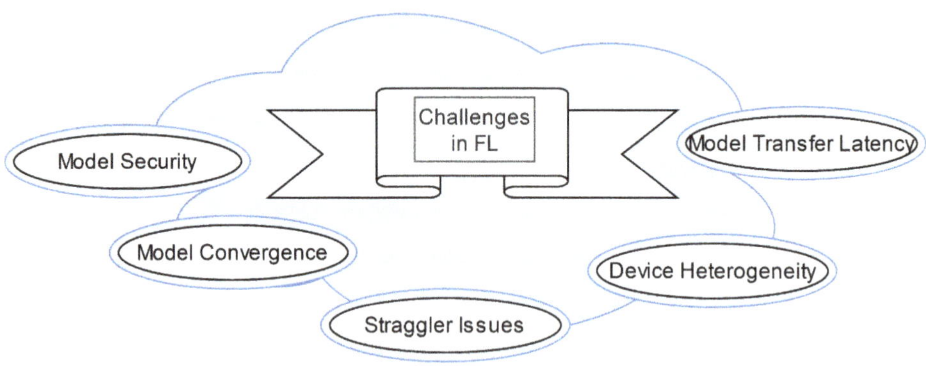

Figure 4.2. Challenges of FL.

strange behaviors while developing an FL model. This is observed in personalized FL frameworks, as studied in [13].

2. Model straggler problem: One of the issues faced by FL frameworks is the model straggler problem. In general, we understand that the models are distributed across multiple hierarchical computing infrastructures that are dispersed across regions. These computing nodes could have varied processing speeds while handling collective intelligence, i.e., updating local models from different edge nodes can have multiple completion times. This situation could lead to a model straggler problem. The problem has been widely studied by several researchers in the past. For instance, the authors of [16] have applied optimization algorithms such as particle swarm optimization (PSO) to mitigate the straggler problems in FL frameworks.
3. Device heterogeneity: The heterogeneity of devices that implement learning models in a distributed fashion can have several issues. For instance, edge-level devices have various power constraints or communication protocols such as WiFi, 5G, or Bluetooth. Similarly, the software packages are not completely loaded on all devices for executing learning algorithms. Accordingly, there will be a substantial amount of delay in the completion of tasks or performance issues when executed on these devices.
4. Model transfer latency: The learning models, being distributed across devices, can experience differing transfer latency. The transfer latency is dependent on communications and the response time of machines. For instance, a Jetson Nano-based edge device could process a portion of the model comparatively faster than Espressif's ESP-based processors. The delay in transferring learning models to a centralized computing machine can hamper the performance of developing models in an FL system.
5. Security challenges: In a federated system, there is a high risk of intruders modifying the models while transferring the findings to a centralized learning system. In addition to several commonly observed network-related security challenges, FL environments face challenges that impact the learning models. A few authors have studied all possible security attacks that could happen in the FL frameworks [8].

4.3 DNN splitting

Edge intelligence can be considered with the concept of split computing. In split computing, the hidden layers of deep learning algorithms are split into portions so that the computations are spread across distributed computing nodes. These computing nodes are often inexpensive when compared to scalable cloud computing units. The authors of [9] explained a few split computing methods that could be applied for deep learning algorithm-based applications.

4.3.1 DNN splitting: architectural concepts

DNN splitting is effective when there are a large number of processing requests from clients. In doing so, the deep neural layers are spread across a larger number of edge

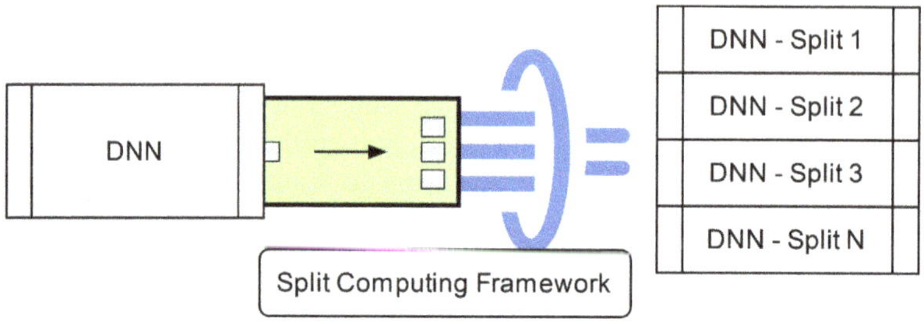

Figure 4.3. Split computing: a conceptual view.

computing nodes than FL frameworks. Obviously, computation, memory, and bandwidth utilization are greater in split computing frameworks than in bare FL frameworks [1].

Figure 4.3 shows the concept of split computing in which DNNs are split into several portions using relevant frameworks.

In fact, the birth of DNN splitting took place due to the difficulties experienced in edge computing systems. The edge nodes are often inexpensive and battery-operated. Many researchers have pinpointed the necessity of edge computing methods instead of clouds for IoT-enabled applications. This is due to the fact that the edge nodes are capable of preventing privacy breach and reducing network traffic. However, executing a large volume of instructions on these edge nodes turned out to be an arduous task. Hence, researchers have designed the concept of DNN splitting, where the computations relating to deep learning algorithms are spread across inexpensive edge nodes [6].

4.3.2 Types of split computing

Split computing is categorized based on identifying the distribution pattern of deep learning algorithms or depending on the time spent deciding the split operations. We understand that split computing has to distribute the computational units within a deep learning algorithm to multiple decentralized computing nodes, including heterogeneous edge-level nodes.

4.3.2.1 Based on distribution pattern

Split computing has to distribute computational units after identifying the portion of computations that need to be spread across the available edge or cloud nodes. Depending on the distribution pattern, split computing is classified into two broad categories:

- Layer-wise distribution: In layer-wise code distribution, the split computing framework has to identify the appropriate cut layer of the deep learning algorithm. The cut layers are either decided based on the availability of a number of edge nodes or the computational capabilities of the nodes. For instance, a 53-layer darknet53 CNN algorithm could be split into several smaller computational chunks considering the computational intensity in each layer and the availability of nodes such as Jetson Nano or scalable cloud nodes.

In general, the cut layers are determined depending on the load requirement of the number of layers and the identified edge node that could perform the computation.

For example, let us assume that a CNN-based deep learning algorithm has to perform layer-specific split computing. A Tensorflow-specific implementation has to load the required modules as follows:

```
import tensorflow as tf
from tensorflow import keras
from tensorflow.keras import layers
from keras.models import Sequential
from keras.layers import Dense
```

Next, after loading datasets, models have to be defined. Refer [2] for the details on CNN algorithms.

```
model_def = models.Sequential([
 layers. Dense(256, activation='relu'),
 layers.Dense(128, activation='relu'),
 …
 …
 layers.Dense(10, activation='softmax')
])
```

In order to perform the split operation, the layers could be divided and distributed to several edge-level nodes as follows:

```
split-level1 = models.Sequential(model.layers[1:2])
split-level2 = models.Sequential(model.layers[2:3])
split-level2 = models.Sequential(model.layers[3:4])
```

- Module-wise distribution: In a module-wise distribution pattern, the layers are not distributed by identifying the cut layers; rather, they are distributed based on identified modules. For instance, a learning algorithm undergoes several stages: (a) importing libraries, (b) loading datasets, (c) pre-processing data, (d) splitting training and testing data, (e) training CNN, and (f) displaying results. Each stage could have several sub-divisions. Notably, CNNs could have several suboperations, such as: (a) defining CNN, (b) adding activation layers, (c) max/avg pooling them, (d) flattening, (e) performing dense layer operations, (f) performing softmax operations, (g) compiling DNN, and (h) displaying results.

  Usually, there are several functional modules that could be executed by multiple edge nodes or computational devices, including cloud servers. If these computational divisions are carried out based on the identified modules, split computing is categorized as module-wise distribution in deep learning.

4.3.3 Dynamic versus static

The cut layers in split computing and the strategy to determine the cut layer can be decided at runtime or prior to executing an application. Depending on the decision time, split computing is categorized into two types:

- Dynamic split computing: Here, the cut layers of a deep learning algorithm are decided while executing an edge-AI application. Typically, researchers decide the layers based on the number of instructions that need to be executed and the computing capability of the underlying hardware. For instance, Jetson Xavier or similar processing boards have sufficient computing capabilities, while ESP32 boards do not have sufficient processing power. According to the knowledge available about the underlying hardware, the split computing framework decides to cut layers and transfer the computational units to the edge-level hardware units. Also, the devices are prone to powering off while performing some actions. Depending on the availability of the computing nodes, the framework could either shrink the number of layers or expand the number of deep learning layers that have to be executed on the compute nodes.

  The major advantages of such decisions are: (i) the flexibility in deciding on the cut layers depending on the availability of nodes, (ii) the capability to speed up computations as they are distributed to powerful edge devices depending on the availability of compute power. However, the drawback of this approach is the huge overhead during the execution of edge-AI applications in deciding on the cut layers and the computational devices.
- Static split computing: Here, the split layers are decided in advance, irrespective of knowledge about the computing nodes. The cut layers of the deep learning algorithm may be determined in an odd fashion or by the number of cuts. For instance, if the 53-CNN-DenseNet algorithm has to be statically split, the computations may be divided into two equal splits, i.e., layers 1 to 26 will be processed on one edge node and the remaining layers will be executed on the other compute nodes.

  As can be seen, static split computing has no dynamism in determining the number of available edge nodes or the capability of the compute nodes for executing the applications. Obviously, the completion time of these applications will be slower, and the resource utilization of the framework will not be as efficient as when compared to the previous approach of split computing.

4.3.4 Advantages of split computing

Split computing has not replaced edge computing. However, the essence of imparting split computing is to benefit from the computational advantages of executing expensive deep learning algorithms on inexpensive computational nodes,

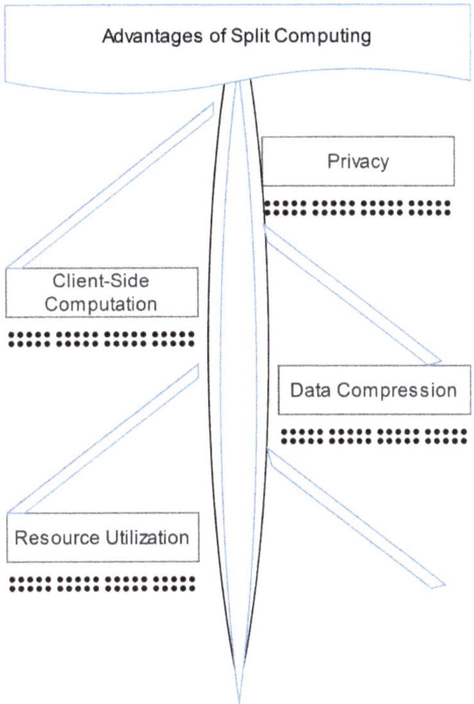

Figure 4.4. Advantages of split computing.

including mobile devices. A few notable advantages of practicing split computing are listed as follows (see figure 4.4):

- Privacy: Split computing shares the common goals of edge computing in increasing the privacy of applications [5]. For instance, video or image processing applications that involve humans are not interested in sharing their pictures with cloud services. They prefer the data to reside on local nodes, such as mobile phones or laptops. One of the major reasons why people do not want their facial images or personal images shared across the internet is to avoid mishandling issues. For instance, internet users could modify online pictures and utilize them to develop funny cartoons or words that narrate fake news or datasets about celebrities. Obviously, it is mandatory to avoid such situations.

  Similarly, in the field of learning analytics, IoT systems are designed to assess the emotions of students to improve their learning capabilities. In such applications, the security could be challenged if the learning algorithms are executed on one single machine at a remote location.
- Client-side computation: The majority of the applications prefer to process clients' requests at the edge rather than in remote locations. One of the major concerns in handling clients' requests on the edge is speeding up the processes involved at the client side.

- Data compression: Data compression is followed in split computing environments while transferring the learning outcomes of intermediate layers. The major objective of including compression is to minimize communication overheads. It increases the network bandwidth of learning layers that are distributed across edge nodes or local computing machines, including mobile phones. Especially when feedback-assisted natural language processing applications or computer vision-related problems are considered, a huge volume of data needs to be transferred between edge–cloud compute nodes or edge–edge compute nodes. It is expected that there will be a large communication delay between edge–cloud compute nodes when compared to edge–edge compute nodes. Hence, developing data compression techniques while performing deep learning analytics can improve the network bandwidth of applications.
- Resource Utilization: Split computing enables all available edge servers to perform operations. Many smaller devices are engaged in processing portions of learning algorithms. In some split computing-based applications, many edge nodes are utilized when compared to cloud servers. For instance, an IoT-enabled learning analytics-based application may engage five to ten cloud servers and over 20 edge servers. The edge servers may be either partially or fully utilized time while executing applications.

  In the majority of the available split computing tools, dashboards are designed to check the utilization of edge servers, local servers, and cloud servers while executing applications. Also, Kubernetes or similar orchestration platforms remain components of component to create a scalable computing environment.

4.3.5 Split computing: tools

There are a few tools that implement the splitting processes of deep neural network layers. The details of SplitFed and SplitLearning are discussed in the following subsections. Although the overall idea of these tools remains the same, the key working functions differ while performing the splitting or organizing of resources to execute the computations on edge nodes.

The tools discussed aim at increasing the training time and mitigating the straggler issues. The training time is increased by splitting the layers across computationally inexpensive edge devices. The stragglers, if any, due to the varying slowness of the devices are avoided by carefully selecting the computational resources.

4.3.5.1 SplitFed

The SplitFed tool [15] was developed with the combination of splitting deep neural networks across decentralized nodes and enhancing the learning processes by applying the concept of FL mechanisms. SplitFed is implemented using the Python language. In the SplitFed tool, the neural networks are partitioned and distributed among client and server nodes. The layer in which the partition is created and distributed to client nodes is termed the *cut* layer. The layers that are distributed across client nodes are actively processed by server nodes in the SplitFed tool, i.e., the results of the client-side neural networks are collected and processed at the server nodes. Accordingly, the server nodes handle a few calculations, such as weighting the results of client-side layers, averaging them, and aggregating the modeling results. Later, the identified weights are transferred to the appropriate clients for further evaluations of *cut* layers.

In the update of the SplitFed tool, the aggregation of the modeling results is not carried out at the server side. Instead, individual clients are responsible for weighting the modeling results and aggregating the learning weights. The advantage of this approach is that it provides less computational load on the server side as servers are busy with other tasks. At the time of writing this book, the framework was developed and demonstrated using ResNet deep neural networks. Further details of the SplitFed tool are discussed in chapter 9.

4.3.5.2 SplitLearning

In general, distributing the neural network layers to distributed computing machines is practiced in several split computing implementations. Almost all the available implementations, such as SplitFed, have dealt with convolutional neural networks. They have not oriented the discussions to models that involve time-series data. For instance, models that involve recurrent neural networks have specific states between learning layers for handling the sequences of data.

In split learning, neural networks that handle sequences of time-series data are partitioned into intermediate layers. At a specific learning layer, the neural network is split and assigned to agents. The agents are responsible for executing the assigned layers and submitting the findings to the servers of the tool. The learning processes involved in the forward propagation and backward propagation steps of the recurrent neural networks are carried out in reverse. In each step, the layers of neural networks are split into intermediate learning layers before they are handled by the learning agents of the SplitLearning tool. Several researchers have utilized the SplitLearning tool to learn time-series data using partitioned recurrent neural network layers.

4.4 Transfer learning

Learning tasks are performed in several forms. One peculiar approach that is historical in nature is transfer learning. In general, transfer learning is the process of learning objects or topics from previously learned tasks. It is a subdivision of the machine learning technique that improves the learning speed of newer tasks if similar tasks already exist.

4.4.1 Traditional view

In fact, the concept of transfer learning is derived from human learning psychology. Children learn a large volume of information from their parents and school teachers. Humans learn topics from civilized society or experienced professionals. In all these learning processes, knowledge has been transferred from one person to another with ease. In doing so, the learner has increased the learning curve more quickly than usual in a subject or domain. For instance, learning to cook pasta or fish curry is propagated over generations by a grandmother and/or mother.

In the context of machine learning, where a human could not learn large volumes of data in varied domains, transfer learning is applied. In transfer learning, models are created for one context and utilized for another learning context. Transfer learning was studied in the 1960s, when AI research started its development. The concept, however, improved in later years with novel algorithms and performance-improvement-enabled solutions.

When compared to machine learning approaches, transfer learning can speed up the performance of learning tasks as it does not need to learn data from scratch to formulate a learning model. This implies that the learning models have to be created independently for each problem domain.

4.4.2 Transfer learning: pictorial representation

Figure 4.5 expresses the concept of transfer learning in a pictorial form. Here, it is easy to understand that learning parameters are extracted by learning models for other related contexts using scalable computing environments such as clouds. However, the edge nodes utilize the features of the learned model for their own target data.

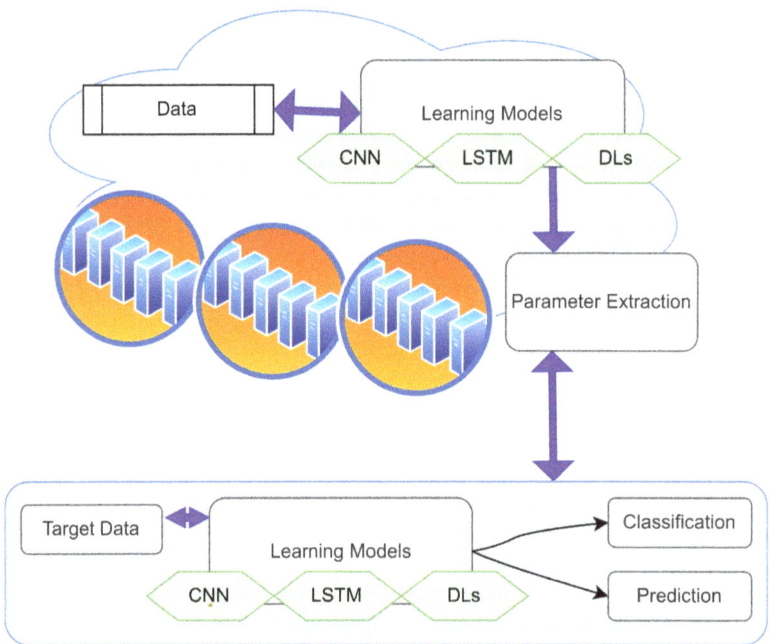

Figure 4.5. Transfer learning: the conceptual diagram.

4.4.3 Transfer learning: stages

Transfer learning utilizes pre-trained models to quickly learn new platforms or datasets. The approach assists tens of thousands of image processing-related applications to quickly accustom learning models to newer data environments. For instance, a learning model created for classifying household objects such as fans, fridges, washing machines, bulbs, etc, could be utilized to train object classification or detection problems. This is due to the fact that most of the shapes were learned by the previously trained model with thousands of pieces of data over a considerable period of time.

The model that is learned using scalable computing nodes and that could be applied to any other learning environment is called the source model, and the learning model that is created utilizing the trained model for newer environments and datasets is called the target model.

The transfer learning process involves several steps to learn data from newer environments. A few notable steps include:

1. Assess data: The data needs to be assessed based on its features and domain expertize. It depends on the conversion of the source model to the target model without much difficulty. While converting the models using the transfer learning mechanism, a few accuracy-related metrics, such as R2, MSE, etc, are utilized in the decision-making processes.
2. Decide learning type: The training data can either be labeled or unlabeled. Accordingly, assessing data varies, and the transfer learning procedure differs, as discussed in subsection 4.4.4.
3. Decide computing node: Transfer learning requires hierarchical computing platforms for faster learning scenarios. At times, the learning platforms have to diligently decide on selecting the computing nodes based on factors such as computational requirements, energy availability, memory availability, etc.
4. Load a pre-trained model: To transfer the learning insights created from one environment to another, a base model is created and loaded. For instance, a model created to detect circles, rectangles, and other shapes could be loaded as a base learning model.
5. Freeze top layers: Freezing the higher layers of the base learning model can enable merging the base model with the newer training models. For instance, freezing the top layers of the CNN algorithm can provide an option to train the base model with the other newer learning layers.
6. Compile models: The last step is to compile the model after merging a few other layering networks with the pre-trained base model. This model can be utilized to infer newer datasets or environments from the pre-trained models.

4.4.4 Transfer learning: types

Depending on which part of the learning model has to be transferred from the source model to the target model, the transfer learning approach can be divided into three different types, as follows:

- Twinned labeled transfer learning: If the labels between the source and target models are duplicated, the learning performed for two different learning datasets is called twinned-labeled transfer learning. In this type of transfer learning, the labels have to be notified across different learning platforms For instance, learning the genders among images captured in different geographical locations falls under twinned-labeled transfer learning. This approach is also called inductive transfer learning.
- Source labeled transfer learning: In this approach, transfer learning is practiced without having labels on the target dataset. The models are designed with labels only on the source model. It is common in many unknown environments to have datasets that cannot be labeled. For instance, an application that identifies animals in a jungle cannot have similar images prior to the investigation for labeling them. Instead, it is required to learn the characteristics of certain wild animals using a few available datasets that are labeled well in advance. In doing so, the rare features of wild animals from a jungle could be detected without any labels at the target locations or images. This type of transfer learning approach is also called 'transducive transfer learning.'
- Unlabeled transfer learning: This type of transfer learning fits well in an unsupervised learning context where there is no possibility of generating authenticated, labeled datasets. The learning algorithm has to create models with the available attributes either by rewarding the best findings or by punishing the wrongly detected findings. Human beings, in general, develop newer theories based on experiences and success stories. For instance, success in a business will be closely monitored by other peer members so that they can follow a similar style of increasing their business profits. Obviously, it may not lead to success in all cases. Accordingly, the success and failures are reasoned out before formulating a success model. In a similar approach, unlabeled transfer learning attempts to transfer the learned model from a source model to a target model without having proper labels in both directions.

4.4.5 Transfer learning: TensorFlow implementation

Transfer learning is widely utilized and implemented in several edge-enabled applications. For instance, the authors of [10] developed a lightweight CNN model to detect COVID-19 patients using learning models developed to detect check infections. Implementing similar transfer learning-based applications is easier using specialized frameworks such as TensorFlow. These frameworks have several in-built features that could be easily integrated with applications using APIs [14]. The base CNN networks have provided many representations based on a large volume of datasets and learning processes in the past. Hence, in the transfer learning process, it is sufficient to merge the last few layers of the base CNN with the newer layers. In doing so, the bare minimal representations of figures and images, such as square shapes, outlines, curves, etc, can be utilized by the new algorithms that merge the transfer learning models.

In Tensorflow frameworks, the following packages are imported to classify flowers or fruits-based on the pre-trained models:

```
import tensorflow as tf
from tensorflow.keras.preprocessing import image
from tensorflow.keras.applications.mobilenet_v2
import MobileNetV2, preprocess_input
from tensorflow.keras.models import Model
```

The mobilenet_v2 is a pre-trained model that can be utilized to classify flowers or fruits. The pre-trained model was created by Google with over 1.4 million images and 1000 classes. It would certainly be a challenging task to develop the learning model from scratch for classifying fruits or flowers. The pre-trained model derived using mobilenet_v2 is considered a base CNN model that could be transferred to computationally inexpensive devices.

Next, the pre-trained model is loaded for the transfer learning processes. The code snippet that is utilized to create such a base CNN model is given below:

```
base_model    =    MobileNetV2(input_shape=SHAPE-IN-IMAGE,
include_top=False,
 weights='imagenet')
```

As a next step, we need to freeze the layers of the trained base models so that the learned model can be integrated with the newer classifiers or learning models. To freeze the base model, we need to disable the training options of layers using the syntax available in Keras as follows:

```
layer.trainable = False
```

In fact, mobilenet_v2 has many layers that could be frozen from further training processes by setting up the disabling option as mentioned above.

Next, we can apply a few data processing steps, pooling methods, and scaling methods to build the transfer learning models. The inputs of the transfer learning model comprise pre-processing steps and base models to output the classification predictions. For instance,

```
transfer_learning_model    =    Model(inputs=base_model.
input,
outputs=output)
 where,
 inputs = tf.keras.Input() # Followed by pre-processing,
 # averaging, augmenting base_model
 # and, similar new layers
 outputs = Dense(num_classes, activation='softmax')(x)
 model = tf.keras.Model(inputs, outputs)
```

The next step is to compile the model, specifying the metrics, loss functions, and learning rates while optimizing the newer model. Accordingly, the model needs to be trained to create the transfer learning model as follows:

```
model.compile(optimizer=Adam(learning_rate=0.001),
 loss='categorical_crossentropy',
 metrics=['accuracy'])
 history = model.fit(training-data,
 epochs=1000,
 validation-data)
```

4.4.6 Edge-enabled transfer learning

Transfer learning approaches have benefited from the advances in edge computing. In the deep learning era, researchers have applied the concept of transfer learning using edge-enabled devices. For instance, the authors of Abu *et al* [12] involved edge devices to increase the performance of home-enabled healthcare applications. The authors designed deep transfer learning using a labeled dataset to assist patients at home. They utilized a CNN-based pre-trained model on edge devices such as mobile phones to quickly observe the health status of patients.

The application of edge computing in transfer learning not only improves the learning accuracy and latency but also the privacy involved in the learning process. Edge-enabled transfer learning becomes more vital, especially for healthcare-related applications, when the data needs to be transferred to the cloud.

NVIDIA has developed the NVIDIA Transfer Learning Kit to execute pre-trained models on Jetson Nano or similar devices [17]. The kit has support for a large volume of pre-trained models that could be applied in domains such as smart transportation, surveillance activities, etc. A few pre-trained models, such as FaceDetect-IR, TrafficCamNet, VehicleMakeNet, etc, using deep learning algorithms such as DetectNet_v2-ResNet18 or ResNet4, have been widely utilized in the transfer learning research works.

4.5 Gossip learning

In highly distributed data, preserving their privacy and learning the features of the data is a challenging aspect of any machine learning algorithm. The diverse data in a distributed network, however, has to be handled in an elegant manner.

Gossip learning is introduced in the machine learning domain to deal with such situations, especially in P2P distributed networking environments. It is expected that the learning models are developed using gossip protocol principles, i.e., gossip is a P2P approach that expects the nodes to communicate with each other while transferring the state of the information.

Gossip learning algorithms based on deep learning concepts are often scalable owing to their updates on neighboring state information. Also, the probable failures in model creations are vividly known to the neighbors. As data is not transferred across compute nodes, the local data is protected from privacy issues.

The gossip learning approach is further applied in many other applications. For instance, in deep learning algorithms such as CNNs, gossip learning is applied in the optimization stage of the algorithm to converge faster than a traditional approach. In doing so, the gradient descent is quickly optimized, as the model can learn in parallel based on the distributed data.

However, gossip learning has often shown poor performance when compared to several other distributed learning approaches, such as federated learning. In the recent past, authors have explored the performance issues of the gossip learning approach while executing the algorithms on specific communication topologies [3].

4.6 Conclusion

Having learned the base technologies, including operating systems, in the previous chapters, this chapter highlights the importance of utilizing edge devices for developing learning algorithms. It classifies different types of edge-enabled learning algorithms, such as federated learning, split learning, transfer learning, and gossip learning. The chapter provides insights into developing learning algorithms using Tensorflow tools. The ideas expressed in the chapter could motivate readers to develop a similar type of distributed learning algorithm using edge devices.

References

[1] Basu S 2024 Split Learning vs Federated Learning and their Use Cases—piyubasu.medium.com (https://piyubasu.medium.com/split-learning-vs-federated-learning-and-their-use-cases-fe96d5c1394a) (accessed 27 January 2024)

[2] Benedict S 2022 Deep learning technologies for social impact (https://iopscience.iop.org/book/mono/978-0-7503-4024-3.pdf) (accessed 17 February 2024)

[3] Giaretta L and Girdzijauskas S 2019 Gossip learning: off the beaten path *2019 IEEE International Conference on Big Data (Big Data)* pp 1117–24

[4] Guy-Evans O 2023 Occipital Lobe: Definition, Functions, and Location—simplypsychology.org (https://www.simplypsychology.org/occipital-lobe.html) (accessed 27 January 2024)

[5] Jeon J and Kim J 2020 Privacy-sensitive parallel split learning *2020 International Conference on Information Networking (ICOIN)* pp 7–9

[6] Kang W, Chung S, Kim J Y, Lee Y, Lee K, Lee J, Shin K G and Chwa H S 2022 DNN-SAM: split-and-merge DNN execution for real-time object detection *2022 IEEE 28th Real-Time and Embedded Technology and Applications Symposium (RTAS)* pp 160–72

[7] Liu J, Huang J, Zhou Y, Li X, Ji S, Xiong H and Dou D 2022 From distributed machine learning to federated learning: a survey *Knowl. Inf. Syst.* **64** 885–917

[8] Lyu L, Yu H, Ma X, Chen C, Sun L, Zhao J, Yang Q and Yu P S 2022 Privacy and robustness in federated learning: attacks and defenses *IEEE Trans. Neural Netw. Learn. Syst.* https://ieeexplore.ieee.org/document/9945997

[9] Matsubara Y, Levorato M and Restuccia F 2022 Split computing and early exiting for deep learning applications: survey and research challenges *ACM Comput. Surv.* **55** 1–30

[10] Pathak Y, Shukla P K, Tiwari A, Stalin S, Singh S and Shukla P K 2022 Deep transfer learning based classification model for COVID-19 disease *IRBM* **43** 87–92

[11] Shen S, Zhu T, Wu D, Wang W and Zhou W 2022 From distributed machine learning to federated learning: In the view of data privacy and security Concurrency *Comput. Pract. Exp.* **34** e6002

[12] Sufian A, You C and Dong M 2021 A deep transfer learning-based edge computing method for home health monitoring *2021 55th Annual Conference on Information Sciences and Systems (CISS)* pp 1–6

[13] Tan A Z, Yu H, Cui L and Yang Q 2023 Towards personalized federated learning *IEEE Trans. Neural Netw. Learn. Syst.* **34** 9587–603

[14] TensorFlow 2023 TensorFlow Lite API Reference—tensorflow.org (https://www.tensorflow.org/lite/api_docs) (accessed 30 January 2024)

[15] Thapa C, Arachchige P C M, Camtepe S and Sun L 2022 SplitFed: when federated learning meets split learning *Proceedings of the AAAI Conf. on Artificial Intelligence* **vol 36** pp 8485–93

[16] Verma R and Benedict S 2023 Adaptpsofl: adaptive particle swarm optimization-based layer offloading framework for federated learning *Fourth International Conference on Image Processing and Capsule Networks* ed S Shakya, J Manuel, R S Tavares, A Fernández-Caballero and G Papakostas (Singapore: Springer) pp 597–610

[17] Shah C, Wang Y, Aghdasi F and Vishwanath A 2020 Training with Custom Pretrained Models Using the NVIDIA Transfer Learning Toolkit—NVIDIA Technical Blog—developer.nvidia.com (https://developer.nvidia.com/blog/training-custom-pretrained-models-using-tlt/) (accessed 30 January 2024)

IOP Publishing

Edge Intelligence
Deep learning-enabled edge computing
Shajulin Benedict

Chapter 5

Inference/prediction techniques

Edge intelligence includes an inference or a prediction step to reason with data. The edge-enabled learning approaches were studied in the previous chapter. However, the possible approaches to infer data from the trained datasets are equally important for edge-enabled applications as the approaches have to deal with real-time data. This chapter reveals the different inference and prediction techniques that are practiced in edge-AI applications. It suggests different distributed inference-related implementation strategies that are adopted in such edge-AI applications.

5.1 Learning stages

To infer or predict real-time data from the learned model, there are a number of steps to be undertaken in machine learning. They are listed as follows:

1. Data preprocessing: This is the process of shaping data from raw data in various formats. The raw data could be a video, audio, text, or image. Sometimes, the raw data consists of noise data that needs to be either removed or informed to users so that the prediction can be accurately made.

   To do the data preprocessing, certain libraries are readily available in different machine learning-related frameworks. For instance, Python libraries such as 'numpy' filter noise from raw data. Typically, the data is input to the learning system in a CSV file format or text file.

   It is also important to specify the folder where the raw data resides or the location where the real-time data arrives. A few IDEs, such as Anaconda or Spyder, support such stages of data preprocessing in machine learning.
2. Load model: The trained learning model, either in a decentralized fashion or a centralized fashion, has to be loaded to smaller edge nodes considering the space limitations of resources. Sometimes, there are challenges when loading the model such as the non-availability of memory space, compatibility issues, format failures, etc.

3. Variable identification: The purpose of developing machine learning algorithms is to create a mathematical model of the existing real-world problem. To do so, it is a essential to identify the possible number of variables from the pre-processed datasets. Although identifying variables from data is key to creating models, it also contributes in the inference stages for quickly inferring or predicting future instances based on the incoming data and the loaded trained models.

   As we are familiar with the variables, the real-world data in machine learning contexts consists of dependent variables, independent variables, and constants. An independent variable is a variable that contributes to making decisions in a mathematical model. For instance, if the skin texture is a variable defined to determine the age of a person, it is termed an independent variable. This is the variable that needs to be tweaked to study the impact of the output. A dependent variable is a variable that outputs the study from the data. The output could be determined as classification or prediction values. The classification values are bound to have some pre-defined values or classes, whereas the prediction values could have any values. Additionally, there are a few variables that can be multiplied with the independent variables while creating the mathematical models. In this stage of inference, all possible variables need to be extracted from the pre-processed datasets.
4. Data post-processing: Next, the predicted values have to be processed in a user-friendly manner so that the users get more insights into predictions. It is also possible to evaluate the capability of learned models if the results are appropriately processed for ease of use.

Once models are created, inferences or predictions can be made for the data. Inference is utilized to analyze data; prediction is utilized to generate new data. Typically, the inferences are evaluated using specific metrics such as accuracy, false positives, true positives, etc, so that modifications are suggested accordingly in the inference system.

5.2 Inference knowledge levels

The output of learning models based on testing-based real-world data is influenced by several levels of knowledge schemes. Broadly, the learning levels are categorized as (i) descriptive, (ii) diagnostic, (iii) discovery, (iv) predictive, and (v) prescriptive. The inferences obtained from learning algorithms in these five categories are suitable for several applications. Information about these five categories is provided in the following subsections.

5.2.1.1 Descriptive level

In edge-AI applications such as weather reporting or news reporting, there is a need to describe sensed information assessed by some sensors. For instance, based on the sensed data from weather stations such as temperature, humidity, light intensity, etc, it is possible to describe the weather conditions in a geographical region.

At this level, there is no novelty in deriving information from data. However, the descriptive level of inference attempts to describe what has happened in a situation based on the data.

5.2.1.2 Diagnostic level

The knowledge inferred at this level is relatively deeper than at the descriptive level. In some applications, there is an urgent need to assess the reason for the data. For instance, it is important to understand the reason for an edge-AI algorithm to describe the rainy or sunny weather of a day. This means that we need to understand the root cause of the inference rather than describe the data.

However, this level of analytics is not able to suggest novel data or a solution, as the inference is based on the available data.

5.2.1.3 Discovery level

At the discovery level of an edge-AI inference system, novel reasons are identified from a large volume of data. This approach has a deeper insight into the data, typically a huge volume of data. The reasons are generated not only based on the available insights but also based on the newly framed knowledge inferences. Obviously, it has the capability to mention the causes that are not known to many existing stakeholders in the data.

5.2.1.4 Predictive level

In most edge-AI applications, there is a requirement to suggest newer outputs rather than mentioning the root causes of the values or the descriptive information about the sensed information. This level of reasoning can deliver newer output values based on the previous data in the domain. For instance, the weather forecasting application could suggest upcoming weather conditions on an hourly or future-daily basis.

The predictive level of knowledge inference combines historical data and rules described in the learning algorithms to derive future outcomes. Future outcomes often fall into two categories: classification and prediction. This level invites the common question of what would be likely to happen in the near future based on specific prediction algorithms.

5.2.1.5 Prescriptive level

The prescriptive level of inference is more knowledgeable than the previous approaches. In this category, the approach offers the reason for the predictions instead of just providing a few probable future predictions. Obviously, it is better than the predictive level of inferences, as the users are given hints about the reason for the suggested values by machines.

In addition, the system could be developed under this category to suggest multiple decision options with appropriate reasons for choosing them. In this way, users are given more freedom to choose the best available suggestions provided by the machine. The authors of [1] developed a prescriptive method of creating a new system when compared to descriptions while applying the prescriptive learning methods in ML documentation.

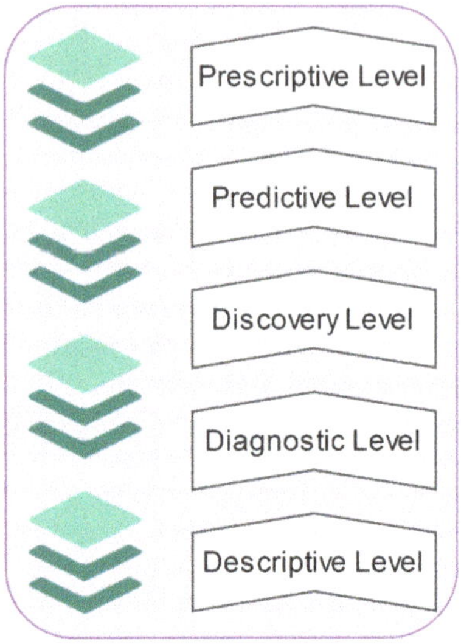

Figure 5.1. Inference levels: knowledge-based distinctions.

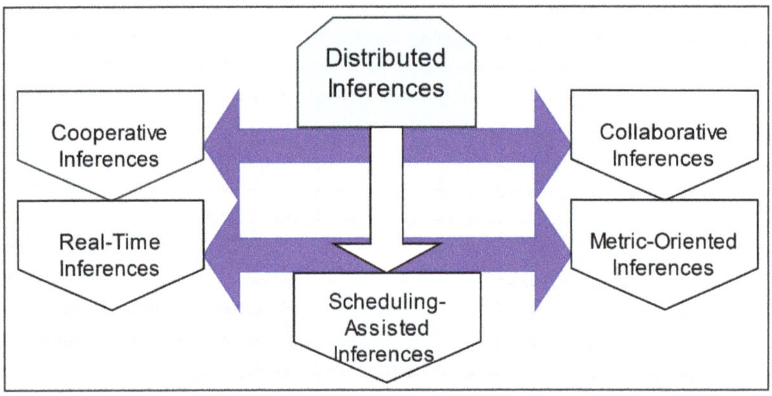

Figure 5.2. Types of distributed inferences.

Figure 5.1 illustrates the crucial learning levels that are distinguished in the inferencing stages of machine learning problems.

5.3 Distributed inferences

Inferences in edge-AI applications are distributed to multiple compute nodes. The compute nodes are geographically spread across regions and continents for some scalable applications. It is a known fact that the data cannot be transferred to cloud or cloud-related servers to perform the learning tasks or inferences. Obviously, distributed inferences are one of the mandates for such edge-AI applications.

In figure 5.2, the probable types of distributed inference techniques practiced by researchers in the past are discussed in detail.

Distributed inferences are often carried out in edge-AI applications with the focus of improving learning accuracy and resource utilization in various ways. The most widely utilized solutions are discussed below.

5.3.1 Cooperative inference technique

The cooperative inference technique utilizes an edge–fog–cloud computing compendium to perform DNN inferences. Typically, the cooperative inference mechanism utilizes a master–slave computing platform such that the learning tasks are distributed across multiple nodes involuntarily based on different learning metrics or performance metrics. The cooperative inference system focuses on splitting the tasks across the participating nodes, considering the lightweight features and heterogeneity of edge nodes. A few partitioning techniques that are developed for distributed inferences are discussed in section 5.6. In [9], the authors have studied the essentials of applying robust communication techniques while devising a cooperative inference mechanism.

5.3.2 Collaborative inference technique

The collaborative inference technique is a voluntary decision of individual participating nodes, including edge nodes, in the process of distributed learning. The collaborative inferences involve edge–cloud. edge–fog, fog–cloud, and edge–edge computing nodes. In some applications, edge nodes contribute to the learning process jointly with a few cloud nodes [5]. This approach has become a common procedure in recent years owing to the requirement to prevent privacy concerns among users.

In the edge–fog collaborative inference mechanism, the devices of edge clusters and fog clusters join together to perform inferences. The involvement of fog clusters accelerates inference decisions nearer to edge-level clusters in a city center or on factory floors. A few acceleration techniques adopted in edge–fog–cloud computing compendium are discussed in section 5.7. Performing collaborative inferences between fog and cloud nodes is not so common. However, the motive for including fog and clouds for inferences is when a large volume of data needs to be processed before making decisions.

Edge–edge collaborations are necessary as learning algorithms are executed on edge-based heterogeneous nodes in many situations. Also, not all edge devices are powered throughout the year when compared to cloud instances. The devices might be voluntarily switched off or be busy doing other user tasks in some situations. Obviously, the ongoing tasks may have to be offloaded to available nodes.

5.3.3 Real-time inference technique

There are a few edge-AI applications that are sensitive to execution time. Also, the inference situations may have to last less than 1 millisecond in applications such as healthcare inference or cybersecurity-related inference tasks.

For instance, an application that needs to observe the emotions of humans and that prevents culprits from criminal endeavors involves several stages:

1. Capturing videos: Videos are captured in real time from different locations using cameras. The hardware units to capture videos have become inexpensive in recent years owing to several technological improvements. The cameras stream data to edge nodes for processing videos. Also, there are a few camera units that include edge processors for filtering and enhancing video data.
2. Framing: Framing is a step in video streaming applications to detect objects or segment objects from images. Framing deals with the aspect ratio of videos, such as wide-screened frames, standard frames, and square-type frames. The typical sizes of these frames are 16:9, 4:3, and 1:1, respectively. Depending on the frame compositions, the quality of the images varies. This is because of the inclusion of objects within frames. For instance, the crime detection application expects that the images resulting from the framing process must include the culprits' information inside frames.
3. Image analysis: It is essential to detect fraudulent images. Also, the cause of crimes. Obviously, deep learning-assisted learning models are required to assess images and crime scenarios. One of the crucial tasks of observing the emotions of humans to detect their criminal endeavors is the facial recognition task. The analysis will be carried out in real-time based on the frames assessed from the captured videos.

5.3.4 Metric-oriented distributed inferences

The distributed inference technique spreads pipelined tasks across compute nodes based on monitored metrics. Edge-AI applications could be represented in the form of data pipelines or workflows. The sub-tasks of such edge-AI applications are often offloaded to distributed nodes. During this process of offloading tasks, it is a better choice to offload them based on measured metrics such as execution time of tasks, energy consumption of tasks, learning accuracy achieved in models, availability of edge clusters, etc. In this way, a fair treatment of inference is achieved with such a distribution of tasks.

5.3.5 Scheduling resources/tasks

One of the few traditional methods that are transported to the edge-AI literature is the application of scheduling algorithms. There are several traditional scheduling algorithms, such as first-come, first-served (FCFS), earliest deadline first (EDF), last-in, first-out, etc. Additionally, there are a few meta-heuristics based on Nature, such as particle swarm optimization (PSO), genetic algorithm (GA), ant colony optimization (ACO), intelligent water drops algorithm (IWDA), honey bee algorithm (HBA), etc.

The main purpose of including scheduling algorithms in inference systems or frameworks for edge-AI applications is to optimally share resources in a collaborative fashion, considering network congestion, energy efficiency of hardware units, resource capabilities, etc.

5.4 Distributed inferences: implementation strategies

Distributed inferences are implemented in several approaches to augment sensors, actuators, and compute nodes. The available implementations excel with the assistance of both software and hardware solutions. In the past, researchers have implemented distributed inferences in three broad categories (see figure 5.3): lightweight implementations, FPGA-based hardware-level implementations, and client–server solutions.

5.4.1 Lightweight implementations

In most cases, edge-AI applications have been developed with a lightweight approach, considering the capabilities of constrained resources. For instance, many lightweight cryptographic algorithms have been developed in the past, considering smaller IoT-enabled edge devices.

A number of researchers have focused on implementing edge-AI applications in a lightweight manner, considering the cost efficiency involved in implementations. To execute DNN algorithms on cloud-based servers, the cost will be extremely high when compared to utilizing the edge-level, inexpensive Raspberry Pi clusters. Obviously, lightweight implementations have boosted the utility of edge-level nodes for executing edge-AI applications.

The lightweight implementations have been introduced in several facets: (a) container approach, (b) microservice approach, and (c) serverless approach.

5.4.1.1 Container-approach

Container-level implementations of servers are lightweight implementations of servers. The implementation of virtualization on top of bare-metal servers using hypervisors has relatively poor scalability owing to heavyweight implementations. In contrast, lightweight containers virtualize server components on top of any host operating system, such as Ubuntu. This enables developers to quickly create containers using two building blocks, such as cgroups and namespaces.

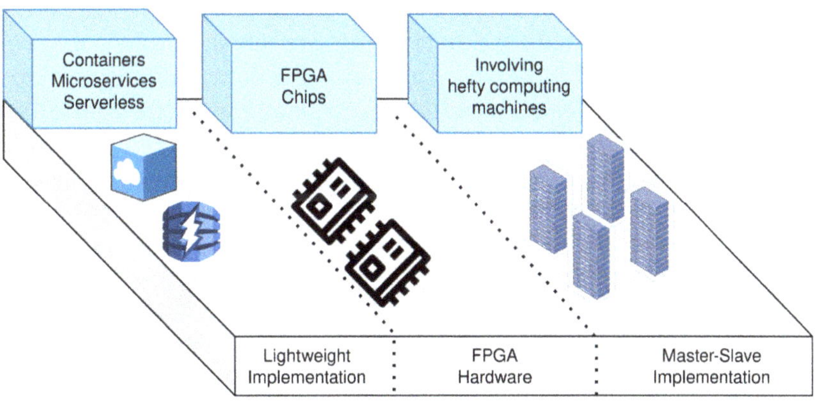

Figure 5.3. Three different implementation strategies in edge-AI.

The cgroups of Linux kernels are responsible for limiting the utilization of underlying resources. For instance, memory cgroups limit the memory utilization of containers to 1 GB or 512 MB based on configurations. Similarly, cpu-set cgroups limit the containers to utilizing only specified CPUs, although the host OS could view all available CPUs.

The namespaces of Linux kernels restrict the views of resource components registered under a container. This feature offers a newer view of containers so that users can feel that they are operating on a newer machine. For instance, namespaces offer new process IDs on newly developed containers starting from PID 1.

The Docker containers are specialized implementations of containers in which they are created considering the union file system of redundant containers, i.e., the containers are not created with all software components such as libraries and kernels if sufficient components are available on the common Docker engine base.

Executing distributed inferences using containers that are hosted on edge-level nodes, including Raspberry Pi clusters or Jetson Nano nodes, is a common practice in several edge-AI applications. A number of researchers have clustered multiple Raspberry Pis using local networks and have hosted inference-related Docker containers on these Docker container-connected pods on top of the Kubernetes cluster for scalable, lightweight implementations.

5.4.1.2 Micro-services approach

Distributed inferences can be implemented in a microservice-based style of implementation. Here, multiple loosely coupled lightweight services could be developed in several languages of choice and discovered during the execution of edge-AI applications. The major advantages of implementing distributed inferences in the microservice architectural style are listed as follows:

1. Distributed development: The microservice-based architectural implementation of distributed inferences enables different developers to independently develop services without any dependency on functional tasks. In contrast, the monolithic services could expect that the teams will have to keep a watch on their peers to understand the updates that are set with a single piece of code.
2. Timely delivery: Due to the functional freedom, the services could be developed independently based on the capabilities of team members. For instance, if the developer specializes in Java, the inference-related services could be developed using Java experts. Also, if another team specializes in Node.js, the tasks could be developed using Node.js experts in parallel. Hence, the time to deliver inference-related tasks will be faster than any monolithic service development.
3. Reusability: Microservice implementations can be repeatedly utilized by multiple applications when compared to other monolithic services. The same implementation can be accessed as individual instances to mark independence for multiple applications.

5.4.1.3 Serverless approach

Apart from microservices, there are efforts to implement distributed inferences of edge-AI applications as serverless instantiations. In general, serverless architecture ensures that the server instances are organized at runtime to execute edge-AI applications. This means that the server instances might be either powered OFF or not assigned to the tasks. Serverless architectures are often maintained by cloud providers. Frameworks such as AWS Lambda, Google Functions, OpenWhisk solutions, etc, have been instrumental in executing serverless tasks.

In recent years, there have been efforts to implement serverless frameworks on edge clusters due to the advantages of the execution environment. Typically, the serverless architecture encourages quick deployments of edge-AI applications without many managerial hurdles. For instance, setting up containers and installing the required software packages has been considered a reasonable overhead for application developers. In serverless architectures, the application developers could focus on implementing their logic rather than setting up overheads. Also, it encourages application developers to stay away from all possible network failure situations that often occur in traditional setups.

Serverless architectures are often combined with event-driven programming frameworks. This feature drastically reduces the cost involved in operating edge-AI applications. The event-driven programming has options to trigger the instantiation of server instances on edge clusters or power ON edge devices to learn sensed data. It could also be utilized for triggering instances for inferring data. In this way, events are triggered based on sensor data or actuator signals that are not very frequent in several edge-AI applications.

5.4.2 FPGA-based implementation

FPGAs have been utilized in edge-AI applications to accelerate the speed of computations and inferences with limited energy consumption. This is due to the fact that independent execution of instructions or function modules is steered in FPGA-enabled edge-AI applications. The FPGAs could integrate memory, CPU, and the other required I/O chipsets on a single device. They are often implemented as a single device unit that mounts all relevant components.

In fact, there are several stages of FPGA implementation. This includes hardware customization, software mapping, and execution. Customizing hardware setup is one of the crucial steps in FPGA architectures, as the hardware units could be configured so that data inferences are wired to customized application domains. Also, software mapping is considered an equally important feature of hardware customization due to the necessity of achieving improved performance. In edge-AI applications, it is important to map learned models to associated hardware units of the FPGA.

The most challenging efforts of FPGA-based implementation of edge-AI applications are to have skillset knowledge in FPGAs. Engineers will require not only hardware knowledge but also a strong skill to understand the edge-AI software and the tuning capabilities of learning models. In fact, this wide knowledge is not very frequently found among modern engineers or student communities. Hence, FPGA-based edge-AI application development is limited to a small community of researchers.

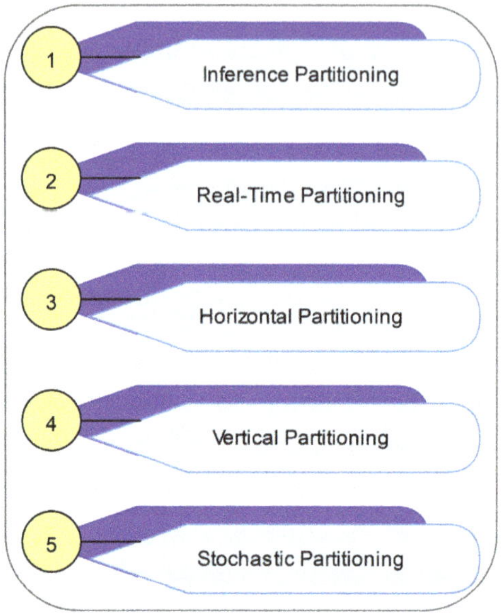

Figure 5.4. Partitioning strategies in edge-AI inferences.

5.4.3 Client–server implementation

A more generic approach to implementing distributed inferences of edge-AI applications is to follow the master–slave approach. The client–server architecture-based implementation has several advantages over the other previous approaches. Notably, they are relatively naive implementation methods. The nodes and software components are easy to manage and deliver tasks as the number of clients and servers is known well in advance in those implementations. The management of clients and servers in a distributed inference system could be controlled unanimously. Obviously, we could effectively deal with the monitoring systems and centrally handle issues, if any.

Figure 5.4 expresses these three different implementation strategies involving different types of hardware for executing edge-AI applications.

5.5 Interactive versus batched inferences

Distributed inferences or predictions are often carried out in two different operating modes, named interactive and batched inferences. These modes represent the outcomes levied by the prediction algorithms based on specific inputs from the applications.

5.5.1.1 Interactive inferences

Inferences that are created with real-time data with one input at a time are termed interactive inferences. The interactive inferences could be integrated into online systems so that users benefit in real-time from the browser features. The inferences are developed based on the requests delivered by users on a demand basis. In some applications, especially when the inferences need to be carried out for individual test data, interactive inferences will be a better solution. However, it can lead to a huge overhead for each prediction.

5.5.1.2 Batched inferences

Batched inferences are experimental modes in which the predictions are carried out for a batch of input data. There are a few applications that do not require an immediate response to the test data. For instance, predicting the sensor information that will be connected to the internet on a weekly basis requires sufficient time to put forth data. In doing so, the offline data could be pre-processed in batches before performing inferences.

The batched inferences, in general, utilize robust big data technologies to handle data while predicting future values. For instance, MPI or Spark programming models could be utilized to deepen the learning inferences more effectively in batches. It can be combined with split-learning frameworks to increase the precision or accuracy of inferences. For instance, the authors of [13] developed a multi-exit learning model that permits batched inferences considering the precision involved in the learning processes.

However, the major drawback of batch inferences is the possibility of a huge waiting time, especially when the data are ready for predictions. Additionally, there is a demand for huge storage space in the batched inference mode of operation.

5.6 Partitioning distributed inferences

Distributed inferences in DNN algorithmic workloads are partitioned to several chunks. The purpose is to increase the utility of machines, improve latency, and scale computations. Figure 5.4 reveals the list of partitioning strategies that are prominently practiced while inferencing edge-AI applications.

One of the major objectives of partitioning inferences is to increase the throughput of inferences. The following subsections provide details of the classified partitioning strategies for edge-AI applications.

5.6.1 DNN inference partitioning

DNN inference partitioning is a naive approach. Here, the DNN inference model is partitioned into smaller computational tasks. The main idea behind partitioning, the inference-related tasks, is to increase the computational speed [3]. Next, the tasks are distributed over heterogeneous computing nodes such as FPGAs, ESPs, accelerators, CPU-based server machines, etc, considering the requirement for computational capabilities. Most of the devices belonging to edge networks are often considered for executing partitioned tasks.

There are numerous strategies followed by researchers in the past while partitioning the inference workloads on heterogeneous edge nodes or compute machines. The important features of these strategies vary based on real-time capabilities, historical perspectives on data, layering workloads, and splitting workloads in a flexible manner.

5.6.2 Real-time partitioning

DNN inference workloads could be partitioned either offline prior to starting the application or in real time. Offline partitioning means that the DNN layers and tasks, including data, that need to be partitioned will be decided well in advance, considering the network bandwidth, heterogeneity of devices, computational capabilities, etc. The partitions are not modified after initiating the execution of tasks. In this way, offline partitioning is not adjustable to varying environments. The partitioning architect or the edge-AI application developer has to keenly distribute the tasks, considering all aspects well in advance.

In contrast, real-time partitioning considers timely decisions while partitioning tasks. The networking bandwidth varies in an edge cluster or the path from edge to cloud; the storage capacity differs over a period of time; the computational permissions or policies of heterogeneous networks vary depending on the node capacities; the energy availability of resources differs on edge nodes; and the budget involved in utilizing computational nodes changes based on the infrastructure providers. Obviously, partitioning needs to be accounted for by all these different factors.

In the past, researchers have developed frameworks and platforms that automate the orchestration of partitioning DNN workloads, including inference workloads, in real time.

5.6.3 Horizontal partitioning

In horizontal partitioning, DNN layers are partitioned horizontally, considering the deeper number of inference layers. The layer in which the partition happens is named the *cut layer*. In the past, practitioners had a single cut layer that divided the inference layers into two partitions. The computations prior to the cut layer and the later stages are independently processed on different networked edge nodes.

5.6.4 Vertical partitioning

In vertical partitioning, the entire inference layer is split into several parts. This leads to a higher level of model parallelism that can be executed on multiple compute nodes, including edge-level nodes. Vertical partitioning is also known as multi-level partitioning.

If the partitioning is leading to several cut layers in the DNN inferences, the mechanism is termed micro-split partitioning. Micro-split partitioning offers smaller computational units for smaller devices that are computationally inexpensive. However, this method requires a faster network in the order of Gbps to fulfill the anticipated transmission delays while transferring computational results between inference layers.

5.6.5 Stochastic partitioning

In a stochastic inference partitioning scheme, the statistical nature of the inference partitioning of layers is considered. Often, the inference processing speed differs

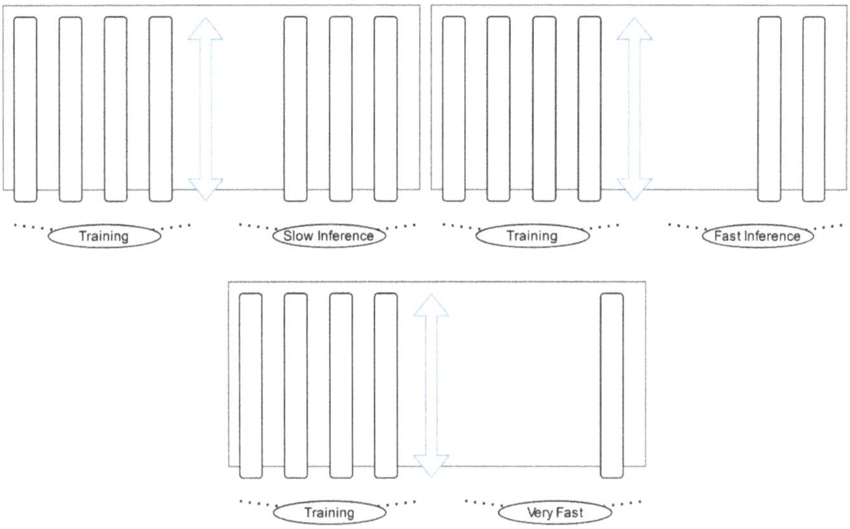

Figure 5.5. Stochastic partitioning of layers in deep learning algorithms.

depending on varying working environments or heterogeneity considerations. Hence, the partitions are distributed across multiple compute machines depending on the inference capabilities, and the stochastic nature of computational capabilities is considered while partitioning the layers.

The number of partitions and the number of layers are dependent on the allocated computing nodes, such as Jetson Nano or other inexpensive processing elements. Figure 5.5 illustrates the concept of slow and fast inferences where the number of training layers differ based on the speed of inferences.

5.7 Accelerating distributed inferences: strategies

DNN inference for edge-AI applications must be accelerated using sophisticated techniques. The major objective of employing acceleration is to increase the performance of inference tasks. In many edge-enabled applications, the acceleration factors depend on appeasing the transmission delays between varying computationally capable devices, i.e., there is a need to balance the fast and slow machines while performing inferences.

Researchers have developed an array of techniques in the recent past to minimize the inference time of applications. Notably, practicing early exit of DNNs, employing edge-level accelerators, and adaptively minimizing the transmission delay between hidden layers are a few techniques adopted in the past. The following subsections explain the approaches that can reduce the inference time of applications by performing a few acceleration strategies.

5.7.1 Hardware and software levels

Many hardware devices have been manufactured in the recent past by Intel, AMD, NVIDIA, and Samsung to accelerate the inference capability of edge-enabled applications. Notably, devices such as ASIC-based Google's TPUs, FPGAs, and GPUs have boosted the realization of involving different constrained devices for inferences. More detailed information about different edge devices and the available accelerators is discussed in chapter 6.

In addition to hardware-level assistance, software solutions have been developed in various approaches to increase the performance of DNN inferences in edge-enabled AI applications. The software solutions are designed as automated tools, libraries, drivers, and middleware.

5.7.1.1 Automated tools

The automated tools are software pieces that can accelerate inferences without any human intervention. The tools are connected with edge-AI applications to offload neural network layers to relevant devices or clusters based on several criteria. Such tools have a few plugins or self-contained software pieces to ease application developers and device engineers.

5.7.1.2 Libraries

These are software-based, non-volatile pieces that can be incorporated into edge-AI applications. The APIs remain as interfaces to these libraries so that the application developers of edge intelligence could utilize them for distributing inference-related tasks to appropriate compute nodes. The libraries are either collaborative or self-contained.

5.7.1.3 Drivers

In some cases, acceleration of DNN inferences needs to tweak the benefits of constrained devices that are not standardized. This means that a few software components are required to influence the operating system of such constrained edge devices to communicate effectively. Accelerators could be fine-tuned by altering the driver features to improve the performance of such applications.

5.7.1.4 Middleware

Middleware-based software solutions will have the capability to identify appropriate accelerators while performing DNN inferences. The middleware connects multiple heterogeneous computing machines by delivering acceleration-related services on top of their operating systems. In fact, middleware components are not available in any accelerators by default. Hence, software solutions that enable such services are termed middleware solutions.

5.7.2 Knowledge distillation

Knowledge distillation is a learning process in which knowledge is transferred from complex learning models to smaller ones. It is an appropriate learning approach for edge-based, constrained devices to accelerate the learning or inference capabilities. In the knowledge distillation approach, the models are distilled or compressed so that smaller devices can train and infer data quicker than the traditional way of making inferences. Knowledge distillation is also termed a teacher–student learning model. The teacher has learned several concepts for many years. But the student collects a few details in an organized manner from the teacher to learn concepts and produce results. This approach to student network learning from the teacher is also termed static knowledge distillation.

The knowledge distillation approach enables lightweight algorithm implementation, which can be executed using lightweight computing machines or battery-operated devices. Another advantage of utilizing the knowledge distillation approach is that it increases the efficiency of robust models.

The authors of [10] proposed a dynamic knowledge distillation approach in which the teacher network constantly updates knowledge based on the input shared by the student network.

5.7.3 Model parallelism approach

One approach that improves the inference time of edge-AI applications is creating a parallel implementation of the inference section of DNN algorithms. In general, a trained model of the DNN algorithm needs to perform inference with novel data to making decisions. Inferring new data from the trained model in edge-enabled machines is complicated due to the consideration of slower machines and battery-operated heterogeneous devices.

The parallel implementations are often implemented in lightweight containers. To minimize the inference time, the DNN models are varied to provide low-speed to high-speed implementations by varying the storage units, learning accuracies, and inference parameters.

5.7.4 Adaptive inference approach

Another approach to improving the inference time of edge-enabled DNN applications is to provide an adaptive approach. For instance, considering the inference time and the prediction accuracy, the execution of layers could be stopped at an early stage. This concept is termed early exiting in DNNs. However, based on the available resources and the time required to predict newer values, the inference tasks could be adaptively increased over a period of time.

In [6], authors have developed an adaptive RESNET algorithm that minimizes the latency and accuracy of the inference system when executed on a resource-constrained edge cluster. The authors have applied a multi-objective optimization step while handling the adaptive inference stage in RESNET algorithms.

A number of researchers have designed an adaptive inference mechanism for DNN where the transmission delay of hidden layers was adaptively modified using computationally powerful accelerators [11]. This approach keeps track of transmitting data between hidden layers that reside on different compute machines on edge nodes. Other researchers tuned the early exit stages of DNNs in an adaptive fashion so that the inferences were accurately handled in the procedure [7].

5.7.5 Model compression

Inferences could be improved in an edge-enabled cloud continuum if large models are effectively transferred from computationally powerful nodes to any neighboring nodes. In general, models that are created in a powerful edge cluster will need a larger storage space and network bandwidth while being transferred to another neighboring edge cluster.

In fact, it is not a good option to generate learning models on all edge clusters for inferences. Instead, transfer learning paves the way to transfer knowledge for easy inferences. However, these learned models have to be compressed to avoid lengthy communication. In [4], the authors have proposed an application of the learning model compression technique using FPGA-based accelerators to reduce the energy and time required for inferences.

5.7.6 Caching inferences

Inferences could be accelerated when they are cached at the first instance. In many edge-enabled IoT applications, the inferences are repeated in a specific pattern. For instance, users may utilize healthcare applications often during a particular period of the day (mostly the morning hours). This reminds the underlying resources, including edge-AI inference engines, to cache the trained model to feed up newer data for inferences. In this way, the cached inferences could deliver better performance for applications.

Caching multiple trained models on an edge server is a challenge. However, there are mechanisms to learn the behavior of input test data patterns to cache models appropriately. The authors of [8] studied the impact of caching models on improving the performance of deep learning inferences. Additionally, the caching approach demands increased storage space on battery-constrained edge nodes.

Similar to caching models, it is possible to cache frequent test data for quick references. Notably, an IoT-enabled edge-based application may require a constant pattern of predicting values. For instance, weather forecasting is an application in which sensor values are collected quite often to make predictions. However, it is not possible to sense and predict values at all geographical locations due to the non-availability of sensors. In such cases, it is enough to cache sensed data from nearer geographical locations for quicker inferences.

5.7.7 Offloading inferences

The algorithmic portion of inferences could be offloaded to either multiple parallel edge nodes or to powerful cloud instances to minimize the computational time of applications. Edge offloading paves the way for caching and acceleration if a framework could automate the offloading processes. For instance, the authors of [2] developed a rewardless guidance algorithm to offload inferences and allocate resources for a large language model.

In literature, offloading is practiced in multiple ways: (a) adaptive offloading, (b) metric-based offloading, and (c) source-aware offloading.

5.7.7.1 Adaptive offloading

Considering the need to complete the inferences or training within a specified time limit, the inference-related tasks are offloaded to clouds or accelerators in an incremental fashion. The adaptive nature of offloading is practiced in many edge-AI applications because of the non-awareness of the constrained device capabilities. For instance, the authors of [12] designed an adaptive offloading framework named FedAdapt while developing a federated learning model and delivering inferences using a few inexpensive resource-constrained edge-level compute nodes.

5.7.7.2 Metric-based offloading

In this approach, a few performance metrics are considered while offloading inference tasks to suitable compute nodes. The most commonly utilized performance metrics considered while deciding on the offloading processes are energy consumption, battery usage, completion time, latency, storage requirements, and bandwidth usage. If the focus is to minimize the energy consumption of inferences, the offloading processes should restrict the computations to slower and fanless edge devices. In contrast, if the completion time is primordial for the offload allocation, computationally powerful resources need to be considered for inferences.

5.7.7.3 Source-aware offloading

In fact, offloading inference-related tasks to networks or edge clusters that do not have sufficient accelerators or that do not satisfy the computational objectives will not solve the purpose. In the source-aware offloading method, the offloads are subject to checking the capability of near-future offloading clusters. For instance, offloading tasks to renewable energy-sourced edge clusters could minimize the carbon emissions of edge-AI-based applications.

5.8 Conclusion

Predicting or classifying data is the outcome of successful learning. Algorithms and techniques that predict data using edge-enabled nodes are specialized topics of research. Accordingly, this chapter explores the distributed inference techniques that could be applied for edge-AI applications. Additionally, a few lightweight implementation techniques involving containers or serverless environments are discussed in detail. The readers can gain insights about the utilization of such techniques for their potential problem domains.

References

[1] Adkins D, Alsallakh B, Cheema A, Kokhlikyan N, McReynolds E, Mishra P, Procope C, Sawruk J, Wang E and Zvyagina P 2022 Prescriptive and descriptive approaches to machine-learning transparency *Extended Abstracts of the 2022 CHI Conference on Human Factors in Computing Systems CHI EA '22* (New York, NY: Association for Computing Machinery)

[2] Fang J, He Y, Yu F R, Li J and Leung V C 2023 Large language models (LLMS) inference offloading and resource allocation in cloud-edge networks: an active inference approach *2023 IEEE 98th Vehicular Technology Conference (VTC2023-Fall)* 1–5

[3] Feltin T, Marcho L, Cordero-Fuertes J-A, Brockners F and Clausen T H 2023 DNN partitioning for inference throughput acceleration at the edge *IEEE Access* **11** 52236–49

[4] Gil-Costa V, Loor F, Molina R, Nardini F M, Perego R and Trani S 2022 Ensemble model compression for fast and energy-efficient ranking on FPGAs *Advances in Information Retrieval* ed M Hagen, S Verberne, C Macdonald, C Seifert, K Balog, K Nørvåg and V Setty (Cham: Springer) pp 260–73

[5] Huang Y, Qiao X, Dustdar S, Zhang J and Li J 2022 Toward decentralized and collaborative deep learning inference for intelligent IoT devices *IEEE Netw.* **36** 59–68

[6] Khan F M, Baccour E, Erbad A and Hamdi M 2023 Adaptive ResNet architecture for distributed inference in resource-constrained IoT systems *2023 International Wireless Communications and Mobile Computing (IWCMC)* pp 1543–9

[7] Laskaridis S, Kouris A and Lane N D 2021 Adaptive inference through early-exit networks: Design, challenges and directions *Proceedings of the 5th International Workshop on Embedded and Mobile Deep Learning* EMDL'21 (New York, NY: Association for Computing Machinery) pp 1–6

[8] Ogden S S, Gilman G R, Walls R J and Guo T 2021 Many models at the edge: scaling deep inference via model-level caching *2021 IEEE International Conference on Autonomic Computing and Self-Organizing Systems (ACSOS)* pp 51–60

[9] Shao J, Mao Y and Zhang J 2023 Task-oriented communication for multidevice cooperative edge inference *IEEE Trans. Wirel. Commun.* **22** 73–87

[10] Tu Z, Liu X and Xiao X 2022 A general dynamic knowledge distillation method for visual analytics *IEEE Trans. Image Process.* **31** 6517–31

[11] Wang H, Cai G, Huang Z and Dong F 2019 ADDA: adaptive distributed DNN inference acceleration in edge computing environment *2019 IEEE 25th International Conference on Parallel and Distributed Systems (ICPADS)* pp 438–45

[12] Wu D, Ullah R, Harvey P, Kilpatrick P, Spence I and Varghese B 2022 FedAdapt: adaptive offloading for IoT devices in federated learning *IEEE Internet Things J.* **9** 20889–901

[13] Zhang S, Cui W, Chen Q, Zhang Z, Guan Y, Leng J, Li C and Guo M 2022 PAME: precision-aware multi-exit DNN serving for reducing latencies of batched inferences *Proceedings of the 36th ACM International Conference on Supercomputing* ICS '22 (New York, NY: Association for Computing Machinery)

IOP Publishing

Edge Intelligence
Deep learning-enabled edge computing
Shajulin Benedict

Chapter 6

Edge resources and accelerators

Edge computing has become a niche area between sensors and actuators and centralized cloud environments. It increases the computational capability of underlying applications. There exist several computing resources with unique characteristics to assist such applications, especially in the AI domain.

This chapter describes the edge-level resources, with more emphasis on edge-AI accelerators. It invites hardware manufacturers to diligently plan the application of varied edge resources for their application domain.

6.1 Edge resources: basics

Microcontrollers and microprocessors are utilized as edge resources while performing edge-level computations in IoT-enabled applications or computer-vision applications. These devices could collect and process sensor information or perform actions through the networked actuators of applications. For instance, Raspberry Pi devices are connected with some connected sensor nodes or actuators to process data nearer to the sensor nodes, a cluster of Raspberry Pi devices is utilized to establish an edge computing device [11], and an array of microcontrollers is utilized as gateways in IoT-enabled applications [5].

6.1.1 Microprocessors

A microprocessor is a digital processing device that consists of processors on a single chip. This processing device is often connected with other peripheral units of edge-connected devices. It is a known fact that microprocessors remain the heart of computers. Similar to computers, microprocessors remain a key processing component of edge computing resources.

The microprocessors of edge resources undergo operations similar to those of traditional computing machines, i.e., they perform computations based on program instructions. They have program counter information that points to the next

doi:10.1088/978-0-7503-5593-3ch6

instruction in a process. In doing so, the program can reside at different memory locations than microcontrollers.

The electronic components of microprocessors have specific devices for performing arithmetic logic operations, encoding and decoding instructions, and fetching operations from correct memory locations without hampering the flow of instructions.

Although the design of microprocessors in edge devices remains the same, they have some specific unique features. Notably, these processors are aimed at increasing the scalability and speed of computations. For instance, the 4th generation's Intel Xeon processor is designed to accelerate the video processing capabilities of edge-related applications.

6.1.2 Microcontrollers

Microcontrollers have been in existence for decades. Computer architects are, in general, interested in building embedded applications using microcontrollers rather than microprocessors. The main reason behind this is the advantages of microcontrollers for specific applications.

In general, microcontrollers embed computing components, I/O units, and memory within a single chip or integrated circuit. The microcontrollers are compact enough to accommodate the required number of memory elements and I/O peripherals within the chip. Hence, such electronic devices are capable of processing customized programs using their components.

The majority of the microcontrollers utilized in sensor-based applications have additional components such as analog-to-digital converters, digital-to-analog converters, connecting buses, and I/O ports. These individual components of microcontrollers have been designed to assist specialized sensors or actuators that are attached to the system.

Often, the processors in microcontrollers are designed using reduced instruction-set architectures (RISC) rather than complex instruction-set architectures. For instance, Espressif's ESP32-based systems, such as ESP32 C6, ESP32 C3, or ESP32, are based on RISC-V architecture. However, CISC-based architectures could be utilized for designing microcontrollers.

The most crucial features that need to be considered when choosing the best microcontrollers for edge-enabled applications are listed below:

- The devices have to support multiple communication ports.
- The devices must have lower energy consumption.
- The devices built on such microcontrollers should not be expensive.
- The devices should assist developers (both hardware and software developers) in evolving newer designs based on them.

6.1.3 Differences

It is essential to understand the core differences between microcontrollers and microprocessors in edge-enabled systems. The following subsections discuss the major differences among them.

6.1.3.1 Purpose
Microcontrollers are designed for specialized purposes when compared to microprocessors. Often, the memory is limited in microcontroller-based edge nodes. Accordingly, these devices are designed to support specialized processes. In contrast, microprocessor-based devices have unlimited memories as these memory units are organized in a hierarchical fashion on such devices.

6.1.3.2 Circuit capability
Microcontrollers are limited in size and capabilities. As these devices are embedded within a single chip, designing larger circuits is a challenging task. This limitation is often not observed in microprocessor-based edge devices.

6.1.3.3 Peripheral support
Microcontrollers can hold customized peripherals based on the requirements of applications. For instance, an IoT-enabled application consisting of digital sensors could avoid certain converters that increase the power consumption of devices.

6.2 Micro-controller-level devices: an edge?

Having understood the basics of microcontrollers and microprocessors, we need to look into the aspects of applying microcontrollers to process edge-enabled applications. This section explores the necessity of involving microcontroller-based edge nodes.

6.2.1 An edge?

Although edge devices are popularly utilized in several IoT devices, the definition of an edge device was vague at the time of writing this book. A few researchers argue that microcontroller-based devices cannot be classified as edge devices. However, the others represent edge devices as devices that process data nearer to sensor or actuator nodes.

In general, microcontroller-based devices such as Arduino UNO, Arduino Nano, Arduino Due, ESP32, Raspberry Pi Pico, etc could be classified as edge nodes if they could perform real-time processing of data and support multiple communication protocols with less energy consumption. In contrast, the most readily available microcontroller-based devices do not have powerful computational capabilities or the assistance of multiple communication protocols.

6.2.2 Importance of Arduino

Incorporating Arduino boards is very popular in the IoT market. Researchers and practitioners apply Arduino boards for the reasons given below.

6.2.2.1.1 Wider implementation
Arduino boards have wider applicability in various domains such as agriculture, healthcare, finance, education, etc. The industry, which manufactures Arduino boards, has been keen to appease the necessity of users by developing newer boards

based on their requirements. For instance, the nRF52840 product of Arduino has invited edge control solutions using robust remote control features at the IoT gateway; the Yún Rev2 board enables Linux-based execution of applications on edge devices.

6.2.2.1.2 Lower costs

The costs of Arduino boards are much less than those of other edge-based devices such as Jetson Nano or Corel Dev, i.e., the cost of the Arduino UNO board is $3.40 whereas the cost of the Jetson Nano board is $300. This shows that the users are interested in attempting newer practices using Arduino boards when compared to the other edge-level computing nodes.

6.2.2.1.3 Low power

The power consumption of Arduino boards is minimized while designing boards. In general, the power consumption of Arduino boards based on ATMEGA328, ATMEGA32u4, and ATMEGA2560 is 5 volts. The current ranges from 22 mA to 160 mA, depending on the microcontroller boards. The devices could be customized, or the software components could be specifically developed to improve the energy efficiency of applications when executed on these boards.

6.2.2.1.4 Development flexibility

Arduino boards are designed in a modular fashion so that any hardware developer could utilize the bare minimal components to execute their tasks. For instance, if a user requires only two GPIO pins, the board could be customized for the specified requirement. Also, a few analog-to-digital or digital-to-analog converters could be minimized, depending on the required resources. Consequently, the costs involved in the implementation or the energy consumption due to the application could be minimized in Arduino-based hardware implementations.

6.2.2.1.5 Interconnection facility

Arduino boards have provisions to connect with several sensors or actuators available on the market. For instance, it can connect with sensors such as temperature sensors, pressure sensors, humidity sensors, switches, relays, variable resistors, etc. The boards, in addition, have support for connecting with other devices that support varying communication protocols such as WiFi, Bluetooth, LoRAWAN, etc. To do so, the devices are connected with appropriate shields to communicate with these diverse communication protocols. The boards are designed to support varying display units, including digital and analog-based display modes. Also, host machines are allowed to connect with these Arduino boards using the most commonly utilized interface, namely, USB.

6.2.3 Arduino families

Several boards are designed for various purposes, with sufficient libraries and documentation for researchers and practitioners to execute them. Also, the developments have sparked tens of thousands of innovations in the respective fields.

The Arduino boards are often categorized into three broad divisions, as listed below:

1. Nano family: The nano family of the Arduino community has designed boards with mini footprints. These boards have limited support for connecting with all available sensors or actuators due to the minimal power supply or current ranges. The nanoboards have been designed for specific applications such as home automation, agriculture, healthcare, art, industry, etc. The board is designed based on the ATmega328. Nano ESP32 and Nano 33 IoT are a few Arduino boards based on the nano family [13].
2. MKR family: The difficulty with the nanoboards is their ability to handle multiple connectivity protocols such as WiFi, Bluetooth, WAN, etc. As mentioned earlier, the nanoboards were suitable for specific low-cost IoT applications that require limited connectivity support. However, to support multiple connections and establish a wider application spectrum, the MKR family of Arduino boards has been designed.

   In the MKR family, application developers could quickly prototype their ideas due to the availability of inbuilt communication mediums. These boards have built-in connections supporting a range of protocols, such as WiFi, Bluetooth, GSM, and WAN. The authors of [16] have developed a water distribution pipeline assessment system based on IoT platforms that utilize MKR family boards to capture the sensor data.

   However, assistance with varying connectivity protocols is limited for more specific domains, i.e., a board will not support all necessary communication protocols. For instance, narrowband IoT, LoRa connectivity, and SigFox-related protocols are supported by specific boards of the MKR family.

   There are implementations named shields to support a few wider communication protocols in the MKR family of Arduino boards. For example, Arduino MKR ENV Shield Rev2, Arduino MKR GPS Shield, Arduino MKR CAN Shield, etc, have been designed as shields to establish connections.
3. Classic family: The classic family of Arduino boards is designed to support wider applications with sufficient general-purpose I/O pins. The boards based on the classic family are quite popular when compared to the other families. This is due to the fact that the initial practitioners start working with the classic family boards. For instance, an entry-level IoT designer prefers the Arduino UNO board to the boards belonging to the nano or MKR families.

The Arduino boards operate at varying speeds. In order to evaluate the capability of different Arduino boards, a benchmark application, which has 2^9 instructions, was executed on these boards. It was observed that Arduino UNO was relatively slower than the Arduino Due board, i.e., Arduino Due completed the task within 2 seconds when compared to Arduino UNO with 28 seconds. Relatively, Arduino UNO had similar execution times when compared with Leonardo (29 seconds), MiniPro (28 seconds), and Mega2560 (28 seconds). It was also inferred that the task was more than 400 times slower than a normal laptop when executed on these microcontrollers.

6.2.4 Arduino UNO board

The Arduino UNO R3 board is a popular ATMega328-based Arduino board that belongs to the classic family. The board operates at 5 volts, it functions at a 16 MHz operating frequency. The board supports analog and digital inputs and has sufficient pins. Similarly, the board connects with the host machines to flash the edited source code through USB sockets. The board, in addition, has support for transmitting and receiving signals through the inbuilt universal asynchronous transceiver/receiver (UART).

6.2.4.1 Anatomy: Arduino Uno R3

The anatomy of the Arduino Uno R3 board is illustrated with specialized hardware components (pins and sockets): (i) a USB socket, RESET switch, a built-in LED, an ATMega328 microcontroller chip, digital and analog input and output pins, and power sockets.

Figure 6.1 provides a broad picture of the Arduino UNO R3 board with internal components and connections. Although newer board designs could be created using a few major components, initially, users connected sensors or actuators to these readily available boards from Arduino.

6.2.5 Arduino programming

Arduino programs are written in C/C++ using Arduino IDEs. The IDE is specifically written to meet the needs of programmers. The Arduino program

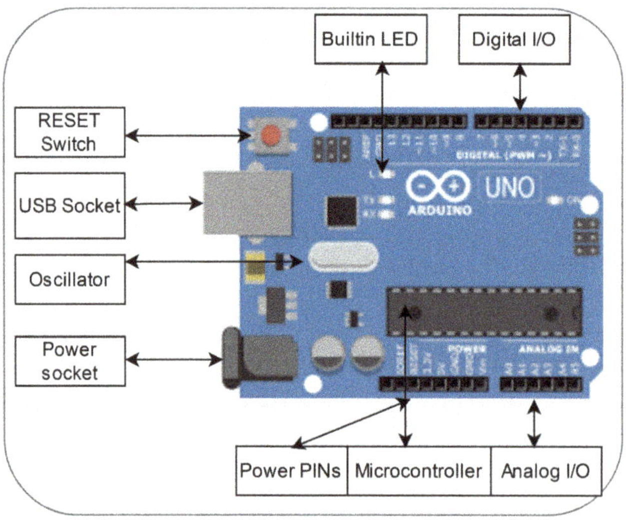

Figure 6.1. Anatomy of Arduino UNO board. Reproduced with permission from Arduino.cc. CC BY SA 3.0.

involves two broadly defined functions, namely setup() and loop() [18]. The setup() function is utilized to configure settings that are necessary for power sensors and actuators. For instance, if there is a need for sending high voltage to a sensor-connected unit through a GPIO pin, the programmer could represent the information in the setup() function. Similarly, continuous measurement-related activities such as monitoring logics from a specific pin number could be carried out using the loop() function, i.e., the loop() function will iterate over the pre-defined tasks continuously without breaks.

The Arduino IDE prepares the environment with ease when considering different boards. Notably, users will have to configure devices with the required power levels and transfer baud rates. The configurations are uniquely organized for different boards. With the utilization of Arduino IDEs, users could easily configure the required settings by selecting the board and the ports. The IDE has an option to view the health status of the user's programs by monitoring them using serial monitors. Also, the IDE has relevant `menu` buttons such as `verify`, `upload`, `new`, `open`, and `save` options to automatically issue commands for flushing, opening, creating, and saving programs to microcontrollers.

6.3 Micro-processor-level devices: general purpose

Typically, edge devices with microprocessor-level architectures are popularly utilized in several applications. The microprocessor-level edge devices include Raspberry Pis, laptops, mobile phones, etc. As mentioned earlier, these devices do not have a single chip containing processors, I/O, memory, and other controllers as they are found in Arduino boards or ESPs.

Edge devices are categorized based on the types of microprocessors. Typically, the microprocessors are classified into CISC or RISC types. CISC is a complex instruction-set computer; RISC is a reduced instruction-set computer. CISC is predominantly utilized in several general-purpose computers, such as ARM-based processors or Intel processors.

The crucial factors that determine a device to be a CISC architecture are: (i) complex instructions that include multiple functions, (ii) variable size of instructions, and (iii) support for various tasks with minimal memory requirements. In contrast, RISC architecture-based microprocessors are meant to improve simplicity in handling resources and relatively fixed-size instructions.

Depending on CISC or RISC architectures, the edge computing nodes are differentiated. The involvement of CISC architectures could increase the complexity of many applications. However, these architectures are applicable in some applications involving virtualization. This enables the edge devices to execute containers and hypervisors, as many instructions could be accommodated in the memory unit of the device when compared to the RISC-based architectures. However, as can be seen, the CISC-based microprocessors could lead to varying CPU cycles for each instruction. This is not the case with the RISC architectures, owing to the simpler instructions. The majority of mobile phones fall under the RISC-based microprocessor architecture, whereas laptops and allied devices resemble the CISC-based edge device.

6.3.1 Raspberry Pi: an example

Raspberry Pi-based devices are the most commonly used edge devices that utilize microprocessor-based architectures. The Raspberry Pi devices support RISC-V architectures [7]. Obviously, the instructions are simpler than on traditional PCs. Also, the execution time of tasks could be easily predicted when compared to the other CISC-based resources.

6.3.1.1 History

In fact, the Raspberry Pi was introduced in early 2006 in the UK by the University of Cambridge. The device was initially developed to enrich school students without any knowledge about edge computing or edge intelligence at this point in time. The main purpose of designing the Raspberry Pi was to make a device that was easily portable by students.

6.3.1.2 Evolution

Raspberry Pi is a small device that can easily fit into a shirt pocket. It could be carried or ported to different geographical locations with ease. It has a 64-bit ARM-based processor. Traditionally, it was applied to learning programming languages such as Python, Java, C, etc. Also, it motivated school students to develop several projects without moving their computers between home and school environments. Gradually, it was identified as a powerful device for developing web servers, gaming solutions, playing HD videos, etc. Recently, with the evolution of edge-enabled AI, the device has become an important component for incorporating learning models and inferring from the learned models.

6.3.1.3 Hardware platforms

Raspberry Pi has been developed at a fast pace as a series of developmental boards to fulfill the requirements of different applications. Notably, there are boards categorized as Raspberry Pi Model A, Model B, Zero, Pi 2 Model B, Pi 4 Model B, Pi 400, etc. Details of these boards are explained below:

- Raspberry Pi Zero: This hardware is a mini-version of the Raspberry Pi. It consists of 40 GPIO pins, a micro-USB port, an HDMI output, etc. The device is compact enough to accommodate a very minimal version of components with a cost of $5 or similar.
- Raspberry Pi Model A: This hardware model is considered to be the traditional model. It was developed in early 2012. These boards had ARM-based processors with a few expensive analog and digital converters. The boards had marked subsequent improvements in various extensions such as A +, B, and pi 2.
- Raspberry Pi 2: This hardware device incorporated a quad-core architecture based on ARM processors. The device is considered to be the fastest pocket-sized single-board computer in 2015. The number of USB ports and the support for Ethernet cables, including HDMI support, will increase in the coming years.

- Raspberry Pi 3: A major drawback of the previous hardware devices is the non-availability of support for other communication protocols such as WiFi, Bluetooth, LoRa, etc. In Raspberry Pi 3, these limitations have been rectified in addition to the traditionally accepted features of boards. This category of boards has high-speed memory and processing power that reaches 1.4 GHz. Additionally, the board supports very high-speed internet using the networking ports. A few variations of this hardware model emerged with specific extensions, such as B and B+.
- Raspberry Pi 4: The Raspberry Pi 4 model is quite popularly utilized in recent years. The main idea behind developing this hardware model is to increase the computational capability of the smaller device. The device has a huge amount of RAM. It targets around 4 GB of RAM, which is similar to a normal laptop. However, the advantage of utilizing such boards is to avoid fans that consume a large amount of power in normal machines. Similar to the other predecessors, this board also features high-end WiFi connectivity, support for multiple communication protocols, gigabit Ethernet, and the other required hardware ports. This particular board operates on 64-bit quad-core ARM processors. A detailed comparison of the most commonly available Raspberry Pi models is described in [4].
- Raspberry Pi 400: One of the most peculiar boards that features the most commonly required computing and networking components in a keyboard is the Raspberry Pi 400. This board is, therefore, termed a keyboard computer. It has integrated the Raspberry Pi 4 boards into the keyboard to ease the portability of devices across geographical regions. Subsequent to the Raspberry Pi 400, there is the emergence of the Raspberry Pi 5 with a very high computing capability and the other facilities incorporated into the board.

6.3.1.4 Anatomy of Raspberry Pi

In general, a Raspberry Pi board consists of a 5 volt USB connector, audio/video connectors, headphone sockets, HDMI ports, USB ports, GPIO pins, Ethernet female connectors, display interfaces, LAN controllers, system on a chip, camera interfaces, and SD card slots. These components are considered the primary components to build a Raspberry Pi board.

Figure 6.2 showcases the anatomy of the Raspberry Pi Model 3 board. The device consists of SoCs, BLE chipsets, GPIO pins, and a few other connectors such as display, LAN, audio/video, etc.

The basic architecture of the Raspberry Pi includes I/O, RAM, CPU/GPU, and a USB hub. The USB hub connects to a few other USB ports, and it has the capability to access the internet through an Ethernet-connected cable. Although all these devices are included in a single board, the chip responsible for computing does not involve other components such as network cards, RAM, or other peripherals.

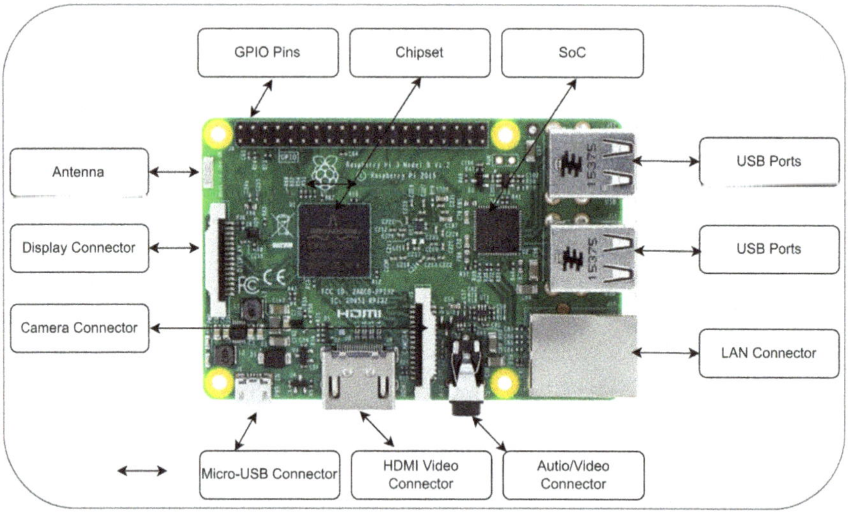

Figure 6.2. Anatomy of Raspberry Pi Model 3 board. Reprinted from Buy a Raspberry Pi 3 Model B+ – Raspberry Pi © Raspberry Pi Ltd. Published under CC BY-ND 4.0 Deed | Attribution-NoDerivs 4.0 International | Creative Commons.

6.3.1.5 Software platforms

The software platforms that are supported by Raspberry Pi boards are programming languages, IDEs, games, and terminal software. The most commonly used programming languages on Raspberry Pi boards are the native Python, C/C++, Java, Nodejs, Golang, and Ruby. These languages are often learned by school students to whom the device was initially recommended.

Officially, the Raspberry Pi is booted with Noobs or Raspbian operating systems. These software platforms are essential software components for organizing processes and monitoring their progress. Additionally, the memory management process is also handled by such system software.

Apart from the traditional OS, there are a few third-party operating systems that are tailored to the requirements of the users. A few operating systems that are popular in this line are Ubuntu Mate, Windows 10 Core, RISC OS, Ubuntu Core, etc.

A few edge-enabled applications that can utilize Raspberry Pi boards are computer-vision-related applications, federated learning-based tasks, collaborative learning platforms, etc.

In general, clusters are created using the Raspberry Pi. Raspberry Pi clusters can be developed using a stack of Raspberry Pi boards. Each Raspberry Pi board is connected using networks. These networks could be tied up to a networking switch, such as a gigabit switch or fiber channels.

Figure 6.3 illustrates the development of a Raspberry Pi cluster using four Raspberry Pi boards. In the figure, we can see that the computing nodes are interconnected using networking cables that are intertwined using port-enabled switches.

In figure 6.3, we can observe that there are four Raspberry Pi boards in a Raspberry Pi cluster. Each Raspberry Pi board is connected to an SD card. The flash

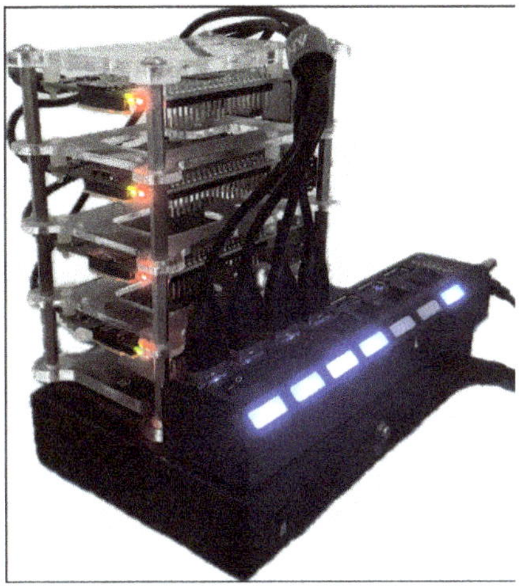

Figure 6.3. Raspberry Pi cluster.

drive's capacity is 64 MB. All five Raspberry Pi boards are interconnected through a dealing-based network switch. This network switch has around six ports.

The main idea for creating such a cluster is to increase the computational capability of applications. We can implement Kubernetes clusters by installing Docker containers and configuring pods on these Raspberry Pi clusters. In general, Kubernetes clusters consist of a master node and several minion nodes. The master node consists of a few software components, such as replica controllers, schedulers, and other monitoring systems. Each Minion node has a few other software components, namely, proxy, Kubelets, and the other user-level monitoring systems.

6.4 GPUs, TPUs, and FPGAs: special purpose

GPUs, TPUs, and FPGAs are special-purpose devices that are designed to cater to the needs of growing edge-enabled applications. The motivation to develop these special-purpose microarchitectures was initiated by a few service-oriented companies, such as Google, NVIDIA, etc.

GPUs are graphic processing units that act as coprocessors for CPUs in a computing machine. They are initially intended to perform computations for processing the output screens. However, researchers identified the importance of such processors while executing single instructions with multiple data points. In general, GPUs utilize CUDA cores to compute instructions for multiple data points in parallel. They follow a unique pattern of handling data in a hierarchical fashion. For instance, there are GPUs that include memory caches at multiple levels to execute instructions offloaded from CPUs. On several GPUs, there are data pipelines and other architectural variations to execute any specialized applications faster, including video and audio streaming applications.

TPUs are tensor processing units that are developed by service-based companies, including Google. TPU microarchitectures are responsible for executing instructions belonging to machine learning domains. Although GPUs and TPUs are considered accelerators, TPUs are more specifically designed based on ASIC technology to appease machine learning or deep learning enthusiasts. As TPUs are more specifically designed for one application domain, they could not be an optimal design for many other applications, including image processing or video processing applications. TPUs are often restricted to a few unique frameworks, such as Tensorflow. However, the core strength of TPUs is the application of the accelerator to several upcoming AI-based edge-level applications.

FPGAs are more specialized and designed for enabling customized solutions. The hardware units are relatively cost-efficient when compared to the other two types of accelerators in edge-level computing systems. The authors of [17] compared the performance of different edge-level accelerators while executing deep learning algorithms.

6.5 SoC, SoM, system-on-board

System on chip (SoC), system on module (SoM), and system on board (SoB) are three chip manufacturing methods for edge-level accelerators. The edge-level accelerators that are designed as the SoC approach include an integrated circuit with all relevant components mounted in it. The idea of integrating all components to improve the acceleration of computing tasks is to minimize the energy consumption of applications. In this way, the battery storage capacity could be significantly increased. In addition, it attempts to reduce the size of the components within a minimal chip footprint. Recently, the authors of [6] have developed a many-core processor-based system on chip that could be utilized for executing deep learning algorithms or for performing edge-level intelligence.

System on module (SoM) is a method of designing edge-level accelerators. When compared to SoC models, SoM has flexibility in choosing components such as CPU, GPU, memory, peripherals, networking ports, sensors, etc. SoM improves the costs involved in designing accelerators. The main reason behind this fact is the limited requirement of resources for applications. For instance, an accelerator might not require sensors in many applications. In such cases, it is better to avoid mounting sensors or actuators on the device.

The most favorable utility of designing accelerators in an SoM approach is to adapt the device based on customized applications. Notably, applications that are specific to increasing security features could feature security-relevant components on SoM. Also, the time-to-market of future applications could be significantly reduced as relevant components could be mounted without delays. Obviously, the manufacturers could keep innovating on the components or portions of components without having a global picture of the hardware design of devices.

System on board (SoB) is a unique approach to manufacturing edge-level accelerators. As we observed, the SoC enabled a chip designer to develop all required hardware components into a single chip; the SoM opened lots of opportunities for a device designer considering various applications and mounting components on top of existing boards. However, in SoB, the entire set of components is attached to a single board, in contrast to the SoM or SoC approaches. It is to be noted that the components could be placed outside a chip. But it has to certainly be laid on top of a single board.

1The major advantage of SoB is that the devices avoid compatibility issues as raised by SoM. This is because the components, such as WiFi, networking ports, CPUs, TPUs, etc, belong to the same proprietary brand as the board manufacturer. In contrast, there is no guarantee of hardware compatibility when the components are designed by multiple vendors. In some cases, the quality of the manufactured components could raise concerns while connecting each other. The disadvantage of SoB is the probable delays in delivering the devices for utility purposes.

6.6 Edge accelerators

Edge accelerators are edge-level devices that intend to increase the processing speed of applications. The characteristics of edge accelerators are listed as low power, compute-intensive, memory-intensive, bandwidth assistive, AI-framework assistive, real-time assistive, and integrative. Learning the characteristics of edge accelerators can enable users or architects to design a novel edge accelerator or utilize edge accelerators in edge-AI applications. Figure 6.4 represents the identified characteristics of edge accelerators, considering devices that are available on the market.

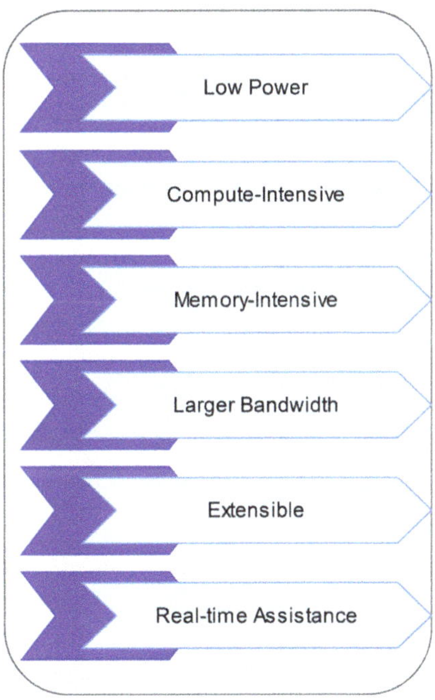

Figure 6.4. Characteristics of edge accelerators.

6.6.1 Low-power

There have been efforts in the past to develop low-power, compact-size edge accelerators that suit multiple application domains. Accelerators such as Kneron, Eyeris, NVIDIA Tegra, AM5729 of Texas, and packaged USB sticks are a few examples of edge-AI accelerators. These devices are designed with minimal power consumption. For instance, Intel's Myriad consumes less than 2 watts [14] when compared to many other edge-AI accelerators.

6.6.2 Compute-intensive

The edge accelerators are aimed at handling as many parallel tasks as possible. They are engaged in tasks such as processing local data, making quick decisions, executing AI-based algorithms considering SIMD principles, etc. In this regard, the top edge-level accelerator manufacturers, such as Intel, Google, and AMD, have focused on executing AI-based applications with high performance. They have designed the hardware to support high-level parallelism.

6.6.3 Memory-intensive

Edge-level accelerators should have the capability to handle higher memory requirements. For instance, the memory size of computer-vision applications or edge-supported IoT applications will be extremely high, in the range of Gigabytes. In order to appease the requirements, the hardware manufacturers, who develop edge-level accelerators, have designed them with sufficient memory size. For instance, Intel has developed Movidius for accelerating computer-vision-related tasks [12] with a memory capacity of 4 GB LPDDR4. In this way, the device can execute deep learning-based inferences without difficulty, even if the applications utilize convolutional networks. The major features of these devices are the non-availability of fans to cool devices, which significantly reduces the power consumption of applications.

Apart from accelerators, there are edge devices with higher memory sizes. For instance, Intel has manufactured the NUC 11 Enterprise Edge Compute NUC11TNHv50L device for executing edge-level applications such that their memory sizes are higher than those of normal machines. Notably, the NUC 11 has around 64 GB of memory with limited computational power. There have been studies to explore the performance of Intel NUC-based devices in the recent past [3].

6.6.4 Improved bandwidth

It is a known fact that bandwidth could be improved in edge-enabled IoT applications or similar federated environments. Efficiently handling the bandwidth requirements of applications is one of the crucial factors in such applications. If not, video streaming or audio streaming can lead to hazardous situations for existing networks owing to the huge rise in network collisions. With the application of edge accelerators, it is expected that the bandwidth will be at least doubled.

Edge clusters have a smaller networking requirement when compared with huge data centers. Such clusters can be utilized for computing instructions where the data resides locally within a restricted networking domain.

Obviously, due to the reduced network bandwidth of applications, the costs involved in the networking essentials will be minimized when the applications are executed on computing machines.

6.6.5 Extensible AI framework

The edge accelerators provide an additional feature for AI-enabled frameworks. In general, AI frameworks offer a set of libraries or APIs to integrate intelligence into applications. These frameworks are built with software modules that are expected to have a higher processing speed. This means that the AI frameworks must be extendable with appropriate proprietary software or non-commercial products. The edge accelerators must have built-in capabilities to boot up, execute, and terminate applications of that sort without much alacrity.

6.6.6 Real-time assistance

The growth of real-time applications and the requirement of solving such applications using real-time capable edge-AI accelerators have become crucial characteristics when choosing hardware. AI designers prefer to include these real-time-capable devices to increase the speed of computations. In fact, there have been several edge-AI applications that are capable of processing data in real-time. Notably, the authors of [9] demonstrated the application of edge-AI in real-time for agricultural farms The authors developed the YOLO-tiny algorithm to detect apples and calculate the potential yield of the farms using edge-level accelerators, including Intel Movidius.

6.7 Commercial edge accelerators

Edge accelerators are manufactured by several industries and research organizations. The most popular edge accelerators are manufactured by the following list of companies: (i) Intel, (ii) NVIDIA, and (iii) Samsung. According to lens.org, the number of patents filed by these companies to October 2023 is given as 1742, 744, and 313 respectively [8].

6.7.1 Intel-based edge accelerators

A few Intel-based products that have been applied in edge-level applications to improve the speed of computations are listed below.

6.7.1.1.1 Intel core processors

Typically, a traditional Intel-core processor could be developed as an accelerator. In fact, such devices will have to have a large number of cores, i.e., a processor in the range of 12 to 24 cores. These processors could be utilized to execute parallel processing tasks effectively when compared to normal server-based compute machines. A few examples of such processors are the Intel i7 series, the i3 series, etc.

6.7.1.1.2 Intel atom processors

In some cases, there is a requirement to optimize the computational speed with respect to power consumption and utilization. For instance, IoT applications relating to computer-vision tasks such as human face identification in airports, malls, etc do not require computations throughout the day. There are specific timings during the night hours in which the processing does not need to be handled. Hence, the accelerators could be adjusted to balance the utilization of power.

6.7.1.1.3 Intel Movidius processors

Movidius processors are exclusively designed for handling vision-related tasks. They have the capability to handle deep learning-related tasks at a higher speed when compared to many others. The idea of developing Movidius-based processors is to enable augmented reality when connected to specialized devices. In fact, Intel's Movidius processor has been well-rated among several researchers and practitioners owing to its smaller size and compatible structures.

6.7.1.1.4 Intel Xeon processors

Intel has been curious about improving the capability of another processor series named the Intel Xeon. The Intel Xeon D processors have been introduced in the market to deliver high performance over other processor models. These processors are capable of handling edge-related tasks using local data sets, resulting in higher performance.

6.7.2 NVIDIA-based edge accelerators

In general, NVIDIA is a company that develops several hardware devices that support CPUs in a SIMD computational model. The purpose of their development is to utilize the GPUs effectively using a few supporting programming languages, such as CUDA or OpenCL. The hardware serves as co-processors for the CPUs when instructions are offloaded to them. In fact, several vector instructions are executed on these hardware units.

The majority of the NVIDIA products, inclined toward edge-level acceleration, fall under three broad categories: (i) core accelerators, (ii) Jetson series accelerators, and (iii) 5G-specific accelerators. The important features of such accelerators for performing edge-level computations are given below.

6.7.2.1 Core accelerators

The core accelerators are the normal NVIDIA GPU cards integrated into edge-based devices such as laptops, workstations, and a few computational nodes. These accelerators are generic ones to compute vectorized instructions for CPUs with higher processing speeds. Typically, these accelerators have over 256 GPU cores to compute data for parallel tasks. Each vector of instructions is executed in one compute cycle when compared to the requirement of n compute cycles in CPUs. A few examples of NVIDIA devices are the NVIDIA Quadro P4000, the NVIDIA Quadro T1000, etc.

6.7.2.2 Jetson series accelerators

With recent developments in the embedded arena, NVIDIA has introduced the Jetson Nano series to improve the energy efficiency of applications. The NVIDIA Jetson Nano series is categorized into the following divisions:

- Jetson AGX Orin series: The Jetson AGX Orin series is a system on chip design architecture [15]. The boards designed for this series deliver around 275 teraflops of computational speed for vectorized instructions. The devices are capable of handling a large volume of data with a memory capacity of over 64 GB. The power of the device could be varied between 15 W and 60 W for the execution of applications.
- Jetson Orin NX series: The Jetson Orin NX series has minimal device space when compared with the other AGX Orin series. However, the computational capability is not equivalent to the previous one. For instance, the device named 'Jetson Orin NX AI Development Kit for Embedded and Edge Systems' computes only 100 teraflops. This is around 2.75x less than the previous series discussed earlier. Also, the memory capacity is limited to 16 GB. The major advantage of this machine is the lower power requirement, which is around 10–25 W, when compared with the previous series.
- Jetson Orin Nano series: The majority of AI developers prefer to develop applications using devices that are not at the production level due to cost factors. Considering the requirements of researchers, NVIDIA has developed NVIDIA Jetson Orin Nano devices that have affordable prices. It can offer over 40 Terra operations of AI instructions, and it can support a vast sector of AI-related applications at the prototype-to-productization levels. With the availability of 8 GB of memory and several ARM-based processors, the device can perform power-efficient computations with ease while developing a prototype solution. However, the Jetson Nano series has a few other models with over 64 GB of memory and SSDs to visualize a production-ready AI-compatible edge accelerator device.
- Jetson AGX Xavier series: The modern Jetson AGX Xavier series model of NVIDIA performs relatively better than a workstation when compared with its size and compute capability. It is developed at a smaller size, nearing 100 x 87 mm. This means that the device could be fixed in any smaller portable device to process locally at the computational speed of 32 terra operations per second. The major advantage of the Xavier series is its capability to process independent data in parallel that corresponds to sensor nodes.
- Jetson TX2 series: The Jetson TX2 series is considered to be the fastest edge accelerator at the time of writing this book. It has been designed with a low-power 7.5 W module to undertake AI-related tasks on edge nodes. Similarly, with the inclusion of over 256 CUDA cores and 8 GB of memory space, the device could accomplish many vector operations within a short span of time.

6.7.3 Samsung-based accelerators

Samsung is regarded as a leading contributor to embedded devices. They have developed embedded solutions to increase edge-level computational capability by providing novelty to memory technologies [10]. The major contributions toward improving the acceleration of edge-level devices are studied in two categories: (i) the processing-in-memory (PIM) category and (ii) the processing-near-memory (PNM) category. The former approach ensures that the computations happen nearer to the data in memory blocks, whereas the latter happens outside the memory blocks. The advantage of the former approach is the avoidance of moving data back and forth between the memory units and the processing elements. However, in the latter approach, a large number of data movements happen between the computational devices and the memory units.

Additionally, Samsung has partnered with other industries to realize the benefits of edge-level accelerators. For instance, Samsung collaborated with Baidu, a Chinese company, to develop an edge-level AI accelerator. The chip named KUNLUN features a 14-nanometer processor with a specialized neural network architecture. The chip can process data at 260 tera operations per second with a power utilization of 150 W. Similarly, it has partnered with AMD Inc., a processor manufacturing company, to incorporate the in-memory technology into AMD processors. The outcome of this collaboration is the birth of a product named the AMD Mi100 accelerator.

6.8 Examples and use-cases

This section explains the application of accelerators for executing edge-enabled applications. The accelerators have been designed for special purposes, such as (i) accelerating the deep learning tasks and (ii) accelerating the audio/video-based streaming tasks.

6.8.1 Accelerators for CNNs

CNNs are convolutional neural networks that are utilized to perform image recognition or detection-related tasks. The CNNs imitate the human brain by recognizing patterns in images. CNNs consist of layers such as the convolutional layer, pooling layer, and fully connected layer. The convolutional layer is responsible for capturing the features of objects in images. Typically, there are several layers of convolutional layers in CNNs. The pooling layer does the process of capturing features of objects by minimizing the number of parameters from them. And the fully connected layer of the CNNs is responsible for classifying objects from images based on activation functions.

The accelerators play a key role in processing CNNs [2]. There have been efforts in the past to study the performance of accelerators when the layers of CNNs were executed on them. Similarly, there have been case studies in the recent past to learn the characteristics of such accelerators while executing CNNs or similar deep learning-related applications.

Strong research that compares the performance of different edge-AI TPU microarchitectures, such as Jetson Nano, and their variants has been studied by the Google research team in the recent past [1]. According to them, the accelerators have marked a very low latency in some versions of such microarchitectures while executing a CNN-based application.

6.8.2 Accelerators for audio

There have been practical implementations that focus on improving the quality of audio perception by involving AI accelerators. For instance, Expedera TimbreAI was developed as an audio inference. This device included an edge-enabled accelerator that operates all relevant audio functions in less than 300 microwatts. The device enables a unified computational architecture to improve the audio quality of applications with very low latency. Also, to improve the performance, this accelerator includes in-memory systems so that the data is processed much faster than audio data. In this way, the accelerator has paved the way for many future audio-based applications [19].

6.9 Conclusion

Learning or inferences utilize edge nodes to perform edge intelligence. However, it is crucial to understand the potential edge-level devices that could be integrated into applications. This chapter reveals a few potential edge-based devices, with more emphasis on microcontrollers such as ESPs and microprocessors such as Raspberry Pis. The importance of utilizing edge accelerators such as Jetson Nano is discussed; additionally, the pros and cons of applying commercial edge accelerators for edge-AI applications are described.

References

[1] Seshadri K K, Yazdanbakhsh A and Akin B 2021 Accelerators for Convolutional Neural Networks—research.google (https://research.google/pubs/an-evaluation-of-edge-tpu-accelerators-for-convolutional-neural-networks/) (accessed 31 January 2024)

[2] Seshadri K K, Yazdanbakhsh A and Akin B 2021 An Evaluation of Edge TPU Accelerators for Convolutional Neural Networks—research.google (https://research.google/pubs/an-evaluation-of-edge-tpu-accelerators-for-convolutional-neural-networks/) (accessed 20 February 2024)

[3] Avari J 2021 Intel NUC 11 Pro (NUC11TNKi5) Review: Tiny Package, Big Potential—gadgets360.com (https://www.gadgets360.com/laptops/reviews/intel-nuc-11-pro-nuc11tnki5-review-india-mini-pc-price-performance-tiger-lake-11th-gen-2502519) (accessed 31 January 2024)

[4] Boards R P 2024 RaspberryPI models comparison — Comparison tables—SocialCompare—socialcompare.com (https://socialcompare.com/en/comparison/raspberrypi-models-comparison) (accessed 31 January 2024)

[5] Chen C-H, Lin M-Y and Liu C-C 2018 Edge computing gateway of the industrial internet of things using multiple collaborative microcontrollers *IEEE Netw.* **32** 24–32

[6] Davies M *et al* 2018 Loihi: a neuromorphic manycore processor with on-chip learning *IEEE Micro* **38** 82–99
[7] Laursen L 2022 RISC-V Guns for Raspberry Pi, Legacy Chips—spectrum.ieee.org (https://spectrum.ieee.org/risc-v-raspberry-pi) (accessed 31 January 2024)
[8] Lens.org 2024 Results The Lens—Free & Open Patent and Scholarly Search—lens.org (https://www.lens.org/lens/search/patent/list?q=Edge
[9] Mazzia V, Khaliq A, Salvetti F and Chiaberge M 2020 Real-time apple detection system using embedded systems with hardware accelerators: an edge AI application *IEEE Access* **8** 9102–14
[10] Samsung Memory 2022 Samsung Electronics Semiconductor Unveils Cutting-edge Memory Technology to Accelerate Next-generation AI | Samsung Semiconductor USA—semiconductor.samsung.com (https://semiconductor.samsung.com/us/news-events/tech-blog/samsung-electronics-semiconductor-unveils-cutting-edge-memory-technology-to-accelerate-next-generation-ai/) (accessed 31 January 2024)
[11] Miori L, Sanin J and Helmer S 2017 A platform for edge computing based on raspberry Pi clusters *Data Analytics* ed A Calì, P Wood, N Martin and A Poulovassilis (Cham: Springer) pp 153–9
[12] Intel Movidius 2024 Intel Vision Accelerator Design with Intel MovidiusVPU—intel.com (https://www.intel.com/content/www/us/en/developer/topic-technology/edge-5g/hardware/vision-accelerator-movidius-vpu.html) (accessed 31 January 2024)
[13] Nano 2024 Nano Family—store.arduino.c.c. (https://store.arduino.cc/pages/nano-family) (accessed 31 January 2024)
[14] Nate Oh 2017 Intel Announces Movidius Myriad X VPU, Featuring Neural Compute Engine—anandtech.com (https://www.anandtech.com/show/11771/intel-announces-movidius-myriad-x-vpu) (accessed 31 January 2024)
[15] NVIDIA Orin 2024 NVIDIA Jetson Orin—nvidia.com (https://www.nvidia.com/en-in/autonomous-machines/embedded-systems/jetson-orin/) (accessed 31 January 2024)
[16] Perez-Padillo J, Puig F, Garcia Morillo J and Montesinos P 2022 IoT platform for failure management in water transmission systems *Expert Syst. Appl.* **199** 116974
[17] Seshadri K, Akin B, Laudon J, Narayanaswami R and Yazdanbakhsh A 2022 An evaluation of edge TPU accelerators for convolutional neural networks *2022 IEEE International Symposium on Workload Characterization (IISWC)* (Los Alamitos, CA: IEEE Computer Society) pp 79–91
[18] Arduino Structure 2024 Arduino Reference—Arduino Reference—arduino.c.c. (https://www.arduino.cc/reference/en/#structure) (accessed 31 January 2024)
[19] TimbreAI 2024 AI Accelerator (NPU) IP—3.2 GOPS for Audio Applications—design-reuse.com (https://www.design-reuse.com/sip/ai-accelerator-npu-ip-3-2-gops-for-audio-applications-ip-50699/) (accessed 31 January 2024)

IOP Publishing

Edge Intelligence
Deep learning-enabled edge computing
Shajulin Benedict

Chapter 7

Performance analysis of edge-enabled applications

One important step to developing an efficient edge-AI application is to diligently plan their executions based on performance. In general, performance studies are a key to improving the development of applications. The method of capturing an appropriate performance metric and planning improvement steps is considered a valid step to improve future developments. This chapter delves into the performance aspects of edge-AI applications and their potential improvement aspects.

7.1 Performance concerns

Edge-AI applications, similar to other scientific applications, are prone to performance concerns. In addition, there are a few design-specific performance concerns in edge-AI systems. In fact, application developers have to look into the capabilities of applications when executed in varying heterogeneous execution environments. In this way, they can improve the quality of edge-AI application execution and the future development of such edge-AI applications.

The performance concerns of edge-AI applications are classified into four broad categories: (a) model-specific, (b) architecture-specific, (c) algorithm-specific, and (d) data-specific. The details of these different categories are explained in the following sections.

7.2 Model-specific performance concerns

Learning models for edge-AI applications are distributed across multiple clusters. Federated environments are often considered while performing the distributed learning. In this regard, there is a potential possibility of model-related performance issues. Some of the reasons why models are prone to performance concerns are listed below:

doi:10.1088/978-0-7503-5593-3ch7

1. The learning models are often partitioned into smaller chunks of computation in edge-AI applications. Notably, DNN models are split into cut layers so that a few layers are executed on specified computing nodes or edge clusters. The partitioning stage in learning models can lead to failures or misconfigured situations on some occasions due to the non-availability of powerful computational nodes at a given point in time. Evidently, this situation raises performance concerns in edge-AI applications. Understanding the requirements and challenges involved in this process, the authors of [12] developed an adaptive split learning framework for learning algorithms.
2. A few learning models are implemented as functions to trigger computational units. At times, the resources are not carefully mapped to the required configurations. The configurational mismatch could lead to performance bottlenecks.
3. The layers of deep learning models are subject to different computational requirements. Hence, training models on homogeneous computing machines or incompatible heterogeneous machines could lead to performance concerns.

7.3 Architecture-specific performance concerns

The performance of edge-AI applications is impacted due to differing architectures (see figure 7.1). Architectures, in general, involve computing machines and the algorithms that drive computations on them. The architectures in edge-AI applications are dependent on the various factors listed below:

1. Hardware-dependent Factors: The hardware-dependent factors relate to designing efficient caches, enabling memory access methods, utilizing appropriate compute resources, etc.
2. Software-dependent Factors: The software-dependent factors suggest the utilization of better operating systems, distributed learning algorithms, and optimally mapped computational algorithms to underlying hardware units.
3. Integration-related Factors: Edge-AI applications require sophisticated communication protocols and integration tools inherent to architectures. For instance, MQTT protocols are sufficient in certain edge-AI applications when compared to applying S-MQTT protocols.

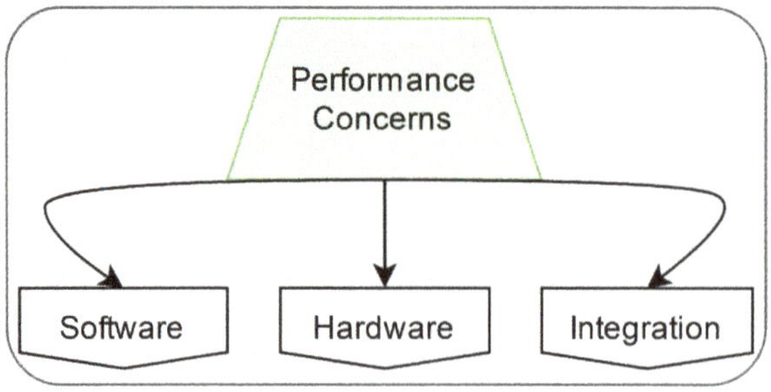

Figure 7.1. Factors impacting edge-AI architectures: broad classification.

7.3.1 Performance concerns: hardware-dependent

In general, computer architecture has followed uniform, non-uniform, and cache-only memory access patterns. The architecture of uniform memory access-based microcontroller or microprocessor devices is designed such that the memory controller can access the corresponding memory units uniformly. Although the access time is uniform, it has poor performance while scaling applications on these devices, i.e., a single memory controller has to deal with all connected memory units.

The advantages of the use of non-uniform memory access-based architectures are that they increase the performance of applications: (a) increasing the performance of applications in near-real-time, (b) supporting a variety of applications with the involvement of multiple memory controllers.

In some applications, we can use cache-only memory access patterns. In cache-only memory access-based architectures, the non-uniform memory access hardware is supported with caches. This means that the memory access is not only based on the location of the memory but also the state and tag associated with the memory units. In fact, the advantages of having different memory access patterns are dependent on the costs involved and space considerations. Notably, cache-only memory access architectures will be much more expensive and complex when compared to simple, uniform memory access-based architectures. The performance, in contrast, offers the opposite results in terms of costs.

The authors of [14] designed a collaborative caching solution to improve the inference time involved in edge-enabled applications. The authors focused on the data involved in the edge-based applications and their caching patterns. The notion was to involve multiple computing nodes at the edge computing level for caching the most frequently utilized data so that the performance of inferences will be faster than with traditional approaches.

Edge-AI applications often need to be ported across computing machines over time. This is because of the distributed nature of computing platforms The hardware-level implementation, therefore, has to involve virtualization techniques for quick portability. For instance, Docker technology delivers an operating system-level virtualization environment where union file systems are maintained to enable a lightweight implementation. It is obvious that Docker-level implementations offer higher portability and self-contained application-related configurations. The authors of [2] developed a lightweight virtualization environment in edge nodes to bridge the gap between cloud and edge-level sensor devices.

In many situations, Docker containers are combined with scalable frameworks such as Kubernetes. The Kubernetes framework can create multiple replicas of Docker instances that work under controlled IP addresses named PODs. The Kubernetes framework has features to orchestrate workloads that are executed on Dockers. The framework could distribute workloads across multiple compute nodes based on the underlying loads. Also, in some situations, there is a requirement for scaling compute instances, especially when faults occur in applications.

7.3.2 Performance concerns: software-dependent

There is no software that fits with all working environments. In recent years, almost all AI-based applications have been moved toward edge nodes rather than processing them in clouds. Having observed such situations, it is crucial to characterize edge-AI applications to satisfy the performance needs of the users. For instance, a video analytic-related application or an audio analytic-related application has to process the streaming data in real-time.

The applications might undergo several stages of execution, such as pre-processing, modeling, prediction, etc. In particular, multiple functions are involved in processing raw data, i.e., functions such as resizing, normalizing, transformations, etc. These functions of streaming applications have to be dealt with within a specified period of time with appropriate hardware devices, including accelerators.

Operating systems in embedded devices or microcontrollers for edge-AI applications contribute to the performance of applications. Notably, this piece of software amalgamation utilizes compute resources rather than allocating all available resources to applications. The authors of [13] compared the performance of two open-source operating systems that are popular in the embedded arena: freeRTOS and ZephyrRTOS.

In general, operating systems aim to increase the utility of resources, monitor the execution of applications, and control the execution of parallel processes on devices such as ESPs or Raspberry Pis.

With the advancement of over 30 years of research, an open-source operating system named FreeRTOS has contributed to the edge-AI domain, including AWS-based edge nodes. FreeRTOS is a suitable operating system that is often utilized in microcontrollers such as ESPs. The operating system has provisions to control or manage resources and offer services to edge-AI applications without much inconvenience to application developers or users.

Without the utilization of operating systems, the applications will not know when to execute them while mapping the available resources at a particular period of time. Additionally, these code snippets could enrich the utility and auditing capabilities of individual applications. A well-written operating system could manage I/O peripherals and sensor/actuator devices while collecting or distributing data. In summary, the software components of the operating system attempt to optimize the performance of the underlying system while running edge-AI applications.

7.3.3 Performance concerns: integration-related

The performance of edge-AI applications is impacted based on the integration of compute devices and storage units in sensor-based environments. The sensors are responsible for sensing data and inducing actuators to perform actions. Depending on the sensor data and the communication protocols, the performance of applications varies. The integration between sensors and actuators through connected devices has to be strengthened in frameworks or integration platforms in order to achieve better performance.

For instance, in industrial IoT platforms, tens of thousands of sensors and actuators are networked through various communication means, including field buses [6]. The integration environments play a key role in determining the automation processes with the utmost reliability or sensitivity. Additionally, the costs involved in the integration of devices while transferring data between edge-AI applications can significantly vary depending on the field buses, such as Modbus, Ethernet Control Automation, Process Field Networks, etc.

One of the major integration issues in any Edge-AI application is the utilization of an appropriate communication protocol. In the application stack, devices often connect to gateways or services through device-level functionalities or wiring topologies. For instance, ESP32 devices can directly connect to cloud-based services or nearby edge-level computing nodes; LoRaWAN devices can connect to gateways before sending data to cloud-assisted services. Also, the bandwidth and power limitations hinder the networking capabilities of these devices. Obviously, the integration of sensors and actuators through varying communication protocols can have a strong impact on the underlying application's performance.

7.4 Algorithm-specific performance concerns

The performance of edge-AI applications can be improved by tuning the algorithms involved in them, apart from hardware recommendations. There are several opportunities, as listed in the following sections.

7.4.1 Compiler-specific refinements

The performance of applications can be improved if the algorithms are compiled specifically for the underlying hardware units. The advantage of involving compilers is to understand the characteristics of the underlying hardware, especially the access patterns of applications. By doing so, the application can be fine-tuned to improve its performance.

A few edge-enabled machines that are more powerful than microcontrollers, such as intel-based fanless devices, have features to select compiler flags considering the optimization features. For instance, flags such as -fO3, -fO2, etc, can induce optimization in application-specific code sequences considering the hardware features of devices, including caches [3]. In distributed environments, compilers are tuned to varying edge-level hardware units.

7.4.2 Tuning applications

Applications can be tuned considering the battery efficiency, brightness level of display devices, switching to power-safe mode, managing sleep modes, turning ON and OFF services, turning ON and OFF hardware devices, automating services, and reducing computations.

Battery efficiency is often achieved by diligently controlling the utilization of embedded devices. Also, the batteries are not charged to their maximum level to achieve a longer battery life. The batteries have to be managed effectively, considering several parameters such as voltage, current utilization, and device utilities.

The batteries, typically, have different characteristics for discharging or charging electrons.

The brightness level of display units on edge devices adjusts the energy consumption of the devices. An increase in brightness levels can have a direct impact on the utility of devices. For instance, in modern devices, ambient-light sensor-based circuits are utilized to automatically vary the brightness of display devices. Accordingly, devices are able to minimize the brightness level of display devices when there are sufficient natural lights.

In general, the power-saving mode increases the lifetime of batteries. In several edge-enabled devices, the batteries are backed up with renewable energy sources. These power sources are often scarce. Obviously, the utilization of power-saving modes in devices could enhance the lifetime efficiency of applications. In some edge-AI applications, the power-saving mode is automatically enacted through APIs, considering parameters such as the availability of power sources.

Increasing the lifetime of applications can indirectly improve their performance. There are a few techniques to improve the lifetime of applications. For instance, enabling sleep modes during the idle time of applications can increase the battery lifetime of devices. The sleep modes are practiced at hierarchical levels. For instance, ESP32 switches OFF WiFi modules at the higher level of sleep mode, whereas the CPU is switched OFF during the deeper instantiating sleep modes. The authors of [9] proposed a solution that switches between ultra-low-power mode and low-power mode depending on the utility of embedded devices.

In a few modern devices, the dynamic voltage-frequency scaling (DVFS) technique is followed to adappropriate to the performance requirements of applications. By increasing the clock frequency of CPUs, the instructions are processed on a larger scale when compared to the reverse situation. Obviously, the energy consumption of applications when executed at a higher clock speed will be higher than otherwise. Other researchers have studied the DVFS government policies for embedded devices, understanding the underlying energy efficiency of the approach. [8].

Additionally, there are opportunities to perform power gating or clock gating techniques on embedded devices to increase the lifetime of applications. In the power gating approach, a few underutilized components are switched OFF; in the clock gating approach, a few underutilized components, often CPU devices, are not provided with clock signals to process the instructions or signals.

Sometimes the performance of applications can be improved by reducing the computation involved in them. For instance, if there is no need to obtain very accurate predictions, it is sufficient to reduce the number of iterations while computing learning models.

7.4.3 Selective packages

Choosing the right package for executing applications can lead to performance improvements. In general, there are several implementations for executing algebraic calculations. However, only a few implementations have mapping configurations

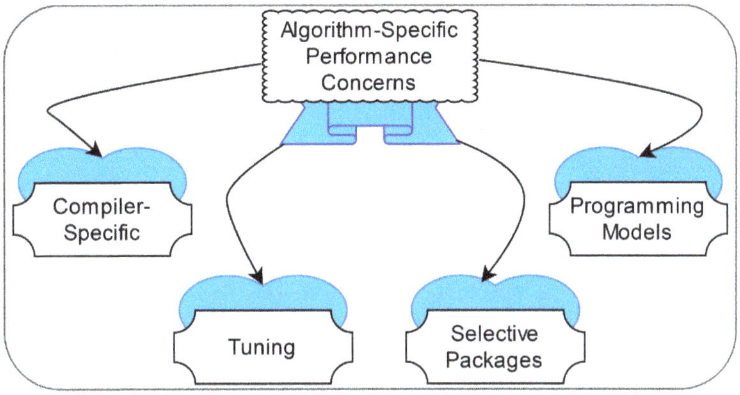

Figure 7.2. Algorithm-specific performance concerns in edge-AI applications.

that fit the hardware units. For instance, the MKL library for intel-based compute machines can have poor performance if the same packages are utilized in ARM or IBM power machines.

Additionally, Node.js-based implementations are sensitive to certain working environments. For instance, npm v1.3.17 does not install dependencies if the same requirements that are specified in the `package.json` file are satisfied. Obviously, the execution of such applications could hamper performance if they are not diligently handled. Similarly, appropriate API references and abstractions are required to handle embedded services such as WIFI, HTTP, logging, networking, etc.

7.4.4 Programming models

The programming model chosen for edge-AI applications will have a direct impact on the underlying performance of logic. Even though business logic could be developed in any programming language, very selective programming models fit well with certain situations.

Big data applications, which need to be processed using files, can be implemented using map-reduce programming models, including Hadoop. Streaming applications that involve audio or video processing can utilize Apache Spark. Also, distributed computing using hybrid programming models such as MPI and OpenMP could fit in specific use-cases. For instance, a fault-tolerant application should not consider an MPI-based programming model while developing its business logic. Obviously, choosing an appropriate programming model will have a major impact on the performance of applications.

Figure 7.2 illustrates the probable performance concerns that are caused by algorithms or application developers who code their applications indifferently.

7.5 Data-specific performance concerns

Data is considered one of the key concerns for improving the performance of edge-AI applications. In edge-AI applications, it is beneficial to have data closer to sensors or actuators. There are several reasons why data has to be closer to users, i.e., privacy and execution requirements have a direct impact while executing applications.

Some approaches in which data can be tweaked to improve the performance of edge-AI applications are listed as follows:

1. Data distribution: If edge-AI applications have options to distribute data across multiple compute nodes, it is better to evenly distribute them across the available nodes. Researchers in the past have studied the impact of the performance of applications while transferring workloads in a dynamic fashion, considering the feedback of executions.
2. Data grouping: It is efficient to pre-process data based on its characteristics before executing the applications. For instance, video-related applications could be pre-processed and grouped so that they could be channeled to fast-moving network channels. in contrast, UDP-related packets could be delivered on highly reliable networks to improve the performance of applications.
3. Data partitioning: In addition to practicing different data distribution patterns, it is also better to partition data, aiming for performance improvements in computing environments. An adaptive nature of partitioning data can pave the way for conglomerating data from multiple applications, considering the limited availability of resources.
4. Quantization: Data quantization attempts to map a series of continuous data points to a smaller set of data points. This reduces the computational time and compute resources required for creating a training model. Developing a robust data quantization technique leverages the performance efficiency of edge-AI applications.

Figure 7.3 shows the performance improvement options that could be adopted in edge-AI frameworks by improving the associated data of applications.

7.6 Performance monitoring: a need

The performance concerns of edge-AI applications have to be captured and rectified. There have been efforts in the past to capture the performance issues of enervated applications. Additionally, careful monitoring is appreciated among users owing to its potential speed up. Monitoring the performance of edge-AI applications, in general, has the advantages discussed below.

7.6.1.1 Bottleneck prediction

It is a known fact that applications or learning models execute well on powerful compute machines. However, transferring data or models across a stringent bandwidth of networks is prone to substantial bottlenecks. Similarly, executing models on low-power, computationally inefficient devices such as the ESP32 can lead to computational bottlenecks due to limited data transfer buses. In some research works, to override such bottlenecks, chip manufacturers have developed a few computing-in-memory systems so that computations are carried out nearer to the data where they reside.

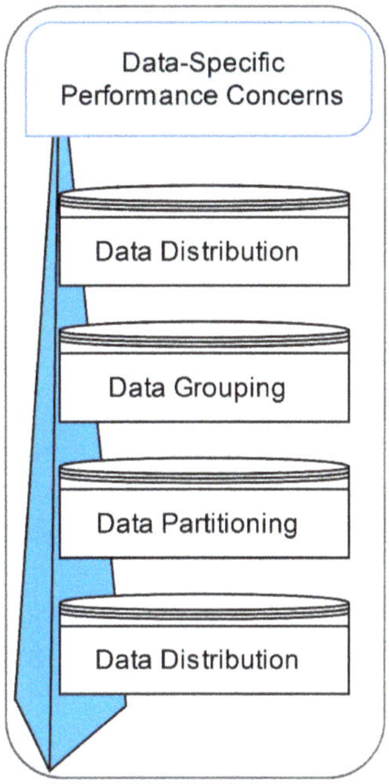

Figure 7.3. Data-specific performance concerns in edge-AI applications.

Obviously, the potential bottlenecks of edge-AI applications are predicted well in advance, and the corresponding applications are tuned to improve their performance.

7.6.1.2 Resource identification

Observing poor performances or behavior of edge-AI applications notifies users and application developers about the cause of the failures or the reasons for success. One of the most challenging aspects of edge-level applications is poor resource usage. Understanding the performance of applications paves the way for selecting appropriate resources in future execution rounds.

7.6.1.3 Visualization

As many resources and compute nodes will be involved in edge-enabled AI applications, it is mandatory to visualize the performance bottlenecks and pinpoint the severity of those bottlenecks. In fact, users or application developers will be engaged in defining logics or utilizing them rather than debugging the issues, if any. Therefore, it is essential to monitor the performance of applications and pinpoint those issues easily.

7.6.1.4 Feedback

Application developers and product developers are keen to understand the user experience and offer suggestions to improve the applications. There is an urgent need to capture those suggestions or comments so that improvements can be made at the right time. Monitoring the performance of applications can directly or indirectly feed inputs to application developers based on such suggestions.

In addition to enhancing the quality of future software releases, offering feedback can reduce the production costs and utility costs associated with the applications. It is necessary to observe that positive feedback often motivates developers to create more innovative solutions within a short period of time.

7.7 Performance monitoring: metrics

Monitoring the performance of edge-AI applications is crucial for improving future application developments. The overall taxonomy representation of the metrics that are utilized to study the performance of edge-AI applications is shown in figure 7.4.

The impact of these performance metrics for assessing edge-AI applications is detailed as follows:

1. Execution time: Execution time is a performance metric that measures the time required for executing edge-AI applications. The applications may be executed on a standalone machine or a combination of multiple compute machines. This metric is essential to evaluating the characteristics of applications based on underlying hardware units. For instance, the authors of [7] studied the efficiency of LSTM algorithms when executed on different

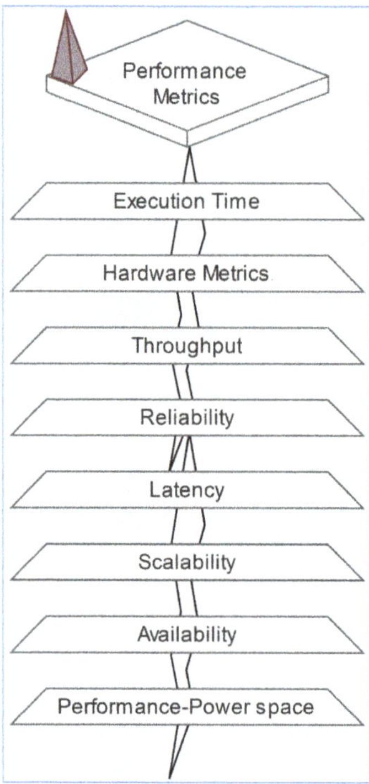

Figure 7.4. Metrics for monitoring the performance of edge-AI applications.

edge accelerators, such as Jetson Nano, based on the execution time of agricultural applications.

2. Hardware-level metrics: Applications, when executed on edge devices, are prone to hardware-related issues, including cache misses or instruction stalls. In all modern computationally capable devices, the hardware metrics could be assessed due to the presence of hardware monitoring counters such as instruction counters. The hardware-level metrics will provide low-level detail about the execution behavior of applications in real-time.
3. Latency: Latency is the performance metric that measures the delay caused in the execution of applications. Latency could be measured due to network delays between edge devices or communication delays between compute components in a machine. The cause of latency is a serious concern for mission-critical, edge-enabled applications. However, in some cases, the applications can be tolerated because of the latency issues. A few researchers have developed a critical view about the latency achieved due to resource-constrained edge devices, i.e., the authors manifested that the latency of applications might become larger in numbers due to the utilization of several resource-constrained edge nodes [1].
4. Throughput or job completion ratio: One of the metrics that determines the capability of the architectural components is termed throughput or the job completion ratio. This metric determines the number of applications or instances executed within a specified time frame. In an edge-AI framework, the throughput could be increased by carefully queuing applications and increasing the number of jobs.
5. Reliability: Reliability is a performance metric that determines the consistency of executing the same application instances over a period of time. In some situations, the applications do not offer the same level of computation for several reasons, including the workload distribution of edge-AI applications across heterogeneous compute nodes.
6. Scalability: Although the performance of applications can be improved by speeding up the computations, the engagement of applications across multiple compute units or increasing the number of application instances is a challenge in several cases. Scalability is a metric that evaluates the importance of scaling applications given a set of compute nodes or frameworks. This metric determines the capability of a framework or architecture for a given edge-AI application.
7. Availability: The availability metric ensures the readiness of compute nodes for computing applications. In most cases, compute resources are not readily available to execute edge-AI applications. This minimizes the performance of applications.
8. Performance–power–space optimization: There is a trade-off between the performance, power, and space requirements of applications. Increasing the performance of applications can speed up their execution. However, the power consumption and utilization of components could significantly increase for the application. Obviously, there is a trade-off between the

performance and energy consumption of applications. Also, in order to increase the performance of applications, there is a tendency to compute applications utilizing multiple compute resources in parallel. This feature can increase the space requirements or memory capacity of executing applications. Hence, in many use-cases, researchers attempt to optimize performance–power–space metrics while executing applications on specified compute nodes.

7.8 Energy-efficiency methods

Energy is a serious concern for battery-operated embedded devices that are utilized in edge-AI applications. This section explores the energy efficiency aspects of edge computing and the research directions in this domain.

In general, power consumption and energy consumption vary based on the application of improvement techniques in the edge-AI domain. Typically, power is considered the multiplication of voltage (V) and current (I). Power consumption relates to the efficiency of the systems involved in executing applications. In contrast, energy is defined as the work performed in a specified time interval. The energy metric directly relates to the time involved in executing an application.

7.8.1 Energy measurements

Measuring the energy consumption of edge-AI applications is one of the most important steps in improving their efficiency. The approach to measuring energy consumption in applications differs from machine to machine. For instance, a computationally powerful edge-level machine that is utilized for executing edge-AI applications could monitor energy consumption using robust modeling techniques. But the energy measurements in a microcontroller-based edge node require instrumenting devices. Accordingly, the energy measurements at edge nodes are classified as discussed below.

7.8.1.1 Energy measurements: data center nodes

The energy consumption of edge devices that are mounted nearer to data center nodes is monitored using intelligent power distribution units. The intelligent power distribution units display the power consumed over a specified period of time. This value is correlated with the execution of edge-AI applications to monitor the energy consumption of applications.

7.8.1.2 Energy measurements: standalone nodes

Several standalone nodes, such as personal computers, laptops, and computationally powerful servers residing nearer to data, are utilized as edge nodes. These devices have in-built sensors to monitor the energy consumption of applications. For instance, Intel-based processors have built-in sensors that model the energy consumption of devices using the running average power limit (RAPL). The energy consumption values that are denoted in joules could be captured for a specified time interval when the applications are executed. There are a few implementations carried out by researchers, such as the PAPI library, to collect the energy consumption of applications using RAPL.

7.8.1.3 Energy measurements: microcontroller-based nodes

Measuring the energy consumption of applications that are executed on microcontroller-based devices is different from other devices. The reasons include the non-availability of hardware counters similar to computationally powerful devices and the non-availability of parallel threads to measure the energy consumption of applications. In addition, any continuous energy monitoring approach could become a huge overhead for such microcontrollers.

To overcome the limitations, the energy consumption of microcontrollers is measured or modeled [10] using external devices that are equipped with current measuring sensors, such as INA219 sensors. This means that a separate measurement device based on another microcontroller device may be required to monitor the energy consumption of applications that are being executed on microcontrollers. These dedicated energy measurement units measure energy consumption using external timestamps that are triggered based on the requirement of a monitoring unit.

7.8.2 Energy efficiency methods

Energy consumption in edge-AI applications is minimized in various ways. Notably, methods such as providing awareness, offering inherent control steps, utilizing appropriate energy sources, and accomplishing algorithm-specific energy efficiency techniques were popularly practiced in the past.

7.8.2.1 Energy awareness

Providing awareness about the energy consumption of edge-AI applications is one step to reducing anticipatory energy consumption. For instance, if an application developer knows that the energy consumption or battery utilization of a specified application is x joules, the developer could attempt to identify the reasons for the energy consumption. Accordingly, energy consumption could be minimized in future developments or implementations.

To do so, the energy consumption of edge-AI applications need to be monitored either at runtime or offline. If the energy consumption of edge-AI application is reported at runtime, the user could either stop the application or modify the runtime configurations to efficiently utilize the application.

7.8.2.2 Inherent control

The energy consumption of applications could be improved if a few control measures are followed. Notably, controlling CPU clock cycles, controlling compiler flags, controlling components, etc, are inherent approaches that improve the energy efficiency of edge-AI applications. The details of the control mechanisms that could be adopted in edge-based devices are listed below:

- Controlling CPU clocks: The clock frequencies are fine-tuned based on the requirements of applications inherently. For instance, edge devices that are computationally capable have adopted dynamic voltage-frequency scaling. With the advent of DVFS, the clocks could be modified to function at varying speeds. Hence, an edge-AI application could be load-balanced, assuming the implications of CPU cycles. The feature of DVFS is inherent in many modern edge-level processors, including Jetson Nano devices.
- Controlling compiler flags: The compiler flags could be tuned to minimize the energy consumption of applications. Flags such as -ffloat-store, -fshort-enums, and -fO3 can significantly modify the code optimizations in edge-enabled learning algorithms. The compiler flags are considered an important optimization tool to deliver energy-efficient applications.
- Controlling components: Switching ON/OFF devices or components in edge-level nodes is a powerful method to minimize the energy consumption of applications. Arduino has features to switch ON/OFF devices, such as analog-to-digital converters, when they are not utilized, using narcoleptic libraries. Also, some software components of applications could be controlled while executing the end-to-end application. For instance, services that are required for monitoring the energy consumption of applications could be stopped if there is no requirement for it to be reported. Similarly, a few security-oriented services, such as identity mapping components, are not required in trusted working environments. Obviously, there are several opportunities to minimize the energy consumption of edge-AI applications by controlling the software and hardware components involved in their execution.

7.8.2.3 Utilizing energy sources

It is an efficient approach to utilize renewable energy sources instead of power grid sources if they are available for power-ON applications. In an edge-AI environment, edge devices could be located in several geographical locations. A few devices may be powered by renewable sources, and the remaining devices may be powered by normal power sources. A framework that is responsible for allocating edge-AI tasks to edge nodes should be diligently allocating tasks to devices that are sourced using renewable energy sources such as solar panels, wind energy sources, or hydropower sources. The advantage of having renewable energy sources is that they minimize the external carbon footprints that are possible using normal power sources, such as thermal power plants.

7.8.2.4 Algorithm-specific energy-efficiency

Edge-AI applications can be tuned to improve their energy efficiency. Carefully developing algorithms that are suitable to be executed in resource-constrained environments is one of the key steps to improving the energy efficiency of applications. In the recent past, researchers have been following several approaches at the algorithmic level to improve their energy efficiency. Some of the solutions are listed below:

1. Datatype selection: The datatype assignment while writing an edge-AI application is diligently carried out considering the requirement of double precision or long integers. There is a tendency for any application developer to initialize a huge stack variable, although the memory requirement might not be the case. In doing so, the memory space and the hardware utility increase without real necessity. This could be easily avoided to improve the energy consumption of such applications. In fact, this method is quite crucial in microcontroller-based edge nodes where memory limitations need to be considered.
2. Thread pinning: Computed resources need to be mapped to the data sources and the availability of data nearer to them. Thread affinity is an approach that ensures the locality of data while executing applications on parallel systems [5]. Algorithms could be designed considering the data affinity factors of the underlying hardware units. Thus, the data need not be transferred across distant devices while performing logic.
3. Handling wait times: In general, operating systems have software modules to control wait times for instructions. The software modules could be interlinked with the application's logic so that the wait time of applications could be minimized in edge-AI applications. In fact, there is a need to constantly assess the wait time of tasks and deliver them to available resources. At times, it is also required to increase the wait times of parallel tasks so that the clock frequencies can be modified to improve the energy efficiency of applications. For example, a parallel task executed on a faster machine ends up with a huge wait time if there are multiple instances of I/O calls. Instead, the clock frequency could be reduced during the wait times so that the energy consumption of the entire application would be minimized.

A holistic picture of the above-mentioned energy efficiency methods in edge-AI systems is given in figure 7.5.

7.9 Carbon efficiency methods

Most researchers are interested in zeroing in on carbon emissions. In fact, IoT-enabled edge-AI applications are prone to emitting a huge volume of carbon due to cooling compute nodes on cloud data centers or on edge-level accelerators. In this regard, tens of thousands of researchers have shown interest in neutralizing carbon emissions [11]. They have attempted to educate the global market about the footprints of carbon emissions. In addition, industries are being trained to prevent carbon emissions by levying policies that are more powerful than in previous decades.

In the past, researchers have worked to remove carbon emissions from industrial waste. To do so, a few researchers have developed real-time working models that showcase the number of carbon emissions from different sources around the globe. The AI-assisted models can pinpoint the forecast volume of carbon emissions while executing edge-AI applications. Accordingly, users could decide to either stop executing such applications or support some policies to override carbon emissions.

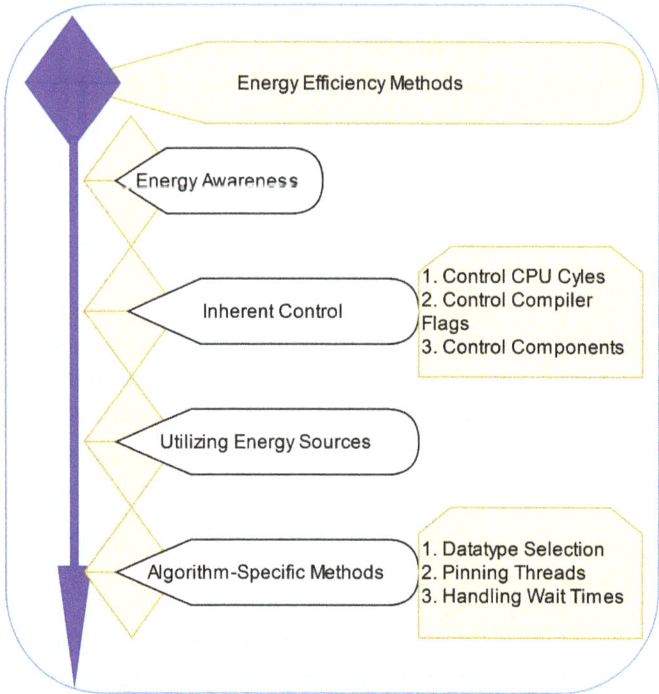

Figure 7.5. Energy efficiency methods.

Carbon emissions will have a direct impact on the performance of applications. To increase the performance of applications, the compute machines are powered ON and processed at a high speed. This attempt can directly heat up the machines and induce cooling systems to indirectly impact carbon emissions in our environment. In a large running edge-AI application, it is sufficient to remove certain execution instances so that carbon emissions can be minimized.

There have been efforts in the past to employ hierarchical computing architectures so that a tree of computing infrastructure could be removed from computations in an edge-AI application. In doing so, carbon emissions could be minimized. Additionally, adopting load-balancing features or data locality concepts could incrementally minimize the carbon emissions of applications.

In some countries, carbon emission policies are finalized so that penalties are framed for defaulters. In order to avoid data immutability, blockchain ledgers are utilized to record the volume of carbon emissions, especially in corrupt industrial locations. As part of carbon emission neutrality, some organizations have attempted to plant trees to neutralize the amount of carbon emissions from their plants. Obviously, data centers that are responsible for emitting carbon emissions will have to plant an equal volume of trees in the name of reforestation to compensate for the carbon emissions. A comprehensive survey of the carbon neutrality procedures adopted by various researchers in the past is discussed in [4].

7.10 Workload scheduling and performance impacts

Scheduling tasks and routing them to different compute nodes in edge-enabled applications have a direct impact on performance. In general, scheduling is a process that allocates resources among the allocated compute nodes. On an edge-cloud platform, this decision is quite cumbersome owing to several factors, such as maximizing resource utilization, minimizing energy consumption, minimizing carbon emissions, minimizing latency, etc.

In edge-AI applications, the most important task is to offload learning models across geographically distributed compute nodes, surpassing from edge to cloud. This is a serious concern for scheduling algorithm designers or policymakers. Due to the ad hoc nature involved in the situation, allocating tasks to the available pool of compute resources is a challenging task.

In the past, researchers have suggested ideas to develop scheduling algorithms, mostly heuristic ones, considering the scheduling objectives. Notably, genetic algorithms, particle swarm optimization, ant volony optimization, and honey bee optimization algorithms have been practiced by a few researchers while deploying workloads across multiple compute nodes of the edge-cloud computing continuum. Apart from the regular meta-heuristic algorithms, there have been efforts to study the impact of the adaptive nature or dynamism involved in the scheduling platforms A few researchers have compared the available scheduling algorithms with the traditional FCFS, or greedy algorithms.

The scheduling decisions are dependent on algorithms that decide the positioning of edge-AI tasks on compute nodes. The scheduling algorithms could establish a schedule such that the tasks are deployed to compute nodes that are powered by renewable energy sources. Accordingly, carbon emissions are minimized in an edge-AI framework. Sufficient research is carried out to look into the best possible scheduling algorithm that can be utilized in any working environment. For instance, a few algorithms are so computationally intensive that they cannot be executed on computationally constrained devices, including ESPs.

Other researchers have explored online scheduling approaches. The online scheduling approaches are highly dynamic in nature. This enables timely decisions considering the dynamism involved in the edge-cloud compendium.

In the existing literature, there are a few limitations in the scheduling algorithms, i.e., the scheduling algorithms have not considered the hidden costs involved in network bandwidth or VM instances utilized for executing the algorithms. In the future, researchers could investigate adaptability or communication-related issues while integrating scheduling algorithms into edge-AI frameworks.

7.11 Performance monitoring tools

Monitoring the performance of applications is an essential step to improving the performance of edge-AI applications. It is essential to understand the performance issues of edge-AI applications and report the issues to the user. This is a mandatory step for application developers to improve the development of edge-AI applications.

In modern situations, monitoring applications are enacted using performance monitoring tools. The tools are necessary due to the following use-cases:

1. It is necessary to showcase the performance problems in a user-friendly manner.
2. The tools are software components that scale automatically along with the applications. For instance, if an edge-AI application is implemented from edge to cloud, the monitoring tool must scale along with the computations of the devices.
3. The performance monitoring tool has to have a limited overhead when compared with applications. This means that the tool itself should not become a burden for the user by executing monitoring-related functions.

In principle, performance monitoring of edge-AI applications could be adopted in five broad categories: the tracing approach, the sandboxing approach, the modeling approach, the simulation approach, and the online approach. The crucial features of these monitoring approaches and their insights are given as follows.

7.11.1 Tracing approach

In the tracing approach, the performance of edge-AI applications is measured after collecting all relevant performance metrics, such as execution time of functions, energy consumption of code snippets, etc. In this approach, the application has to be completed before sniffing over the code portions and the relevant performance monitoring results. The tracing approach is simple due to the measurement pattern that traverses along with the application. In many cases, the approach is also termed an offline monitoring approach.

The major advantage of measuring the performance of applications using a tracing approach is the capability of producing customizable performance analysis reports. The visualization of the performance analysis can be enhanced due to the availability of a large volume of performance data. Additionally, the approach is naive, as there is no need to evaluate the performance of applications in real-time.

However, the major drawback of this approach is the collection of a large volume of performance analysis data. Obviously, a separate processing time is required to study the performance concerns of applications.

7.11.2 Sandboxing approach

Similar to the sandboxes created in playgrounds to ensure a safe playing environment for children, the sandboxing approach in computing systems enables a separate environment to study the performance of edge-AI applications. The approach expects the devices to be compatible with a real-world working environment. The sandboxing environment is an isolated environment that mimics the edge-AI working environment. Hence, users or application developers can learn the probable performance impacts of applications while changing certain runtime parameters, such as the number of compute instances or CPU cycles.

In general, the approach expects a tedious implementation considering various factors of the edge-AI application and the underlying hardware setup. Obviously, the skillset required to develop such a performance analysis environment is difficult in many use-cases. However, once the environment is set up, the sandboxes could provide a more realistic performance detection platform for observing the working behavior of applications.

7.11.3 Modeling approach

The modeling approach expects the performance analysis tool developer to create mathematical models that mimic the edge-AI framework. The modeling approach works based on the prior history and the identification of variables based on the historical data. The historical data suggests variations in the performance of applications due to hardware and allied factors. Accordingly, equations could be modeled so that future experiments are evaluated based on them.

In the modeling approach, the performance of applications is studied after the data is classified into three broad categories, such as training data, validation data, and testing data. The training data consists of performance measures of edge-AI applications that relate to factors, including hardware metrics. Based on the training data, mathematical models are created for assessing the performance of applications. Later, the generated mathematical models are evaluated based on the validation data. During this phase, the mathematical model that mimics the performance model is fine-tuned to attain the utmost clarity. Finally, the testing phase expects that the performance of applications can be evaluated without any further implementations.

The major advantages of this performance analysis approach are the evaluation of the performance of applications without executing them. This is possible based on the mathematical model created. However, the approach expects a large volume of previous historical data to establish an accurate performance prediction model.

7.11.4 Simulation approach

In the simulation approach, the performance analysis of edge-AI applications is derived using simulation patterns. The simulation approach mimics a real-world situation where the developer will have a strong knowledge of the environment. In this approach, users can easily modify parameters to study the performance behavior of edge-AI applications. They can scale certain parameters, such as the number of nodes or communication protocols, without executing applications. As mentioned, developing a simulator is a challenging task due to the requirement of a thorough skillset in hardware and software scenarios.

7.11.5 Online approach

Developing a tool that monitors the performance of applications in real-time is a very challenging task. This is because the overhead of a performance monitoring tool can increase if it is not developed, considering all situations. Typically, the applications are instrumented using specific functions that are invoked at runtime to monitor their performances. For instance, a few functions are developed to mention the start and stop of measurements in applications. These functions establish a stack for measuring appropriate performance metrics such as hardware counters, execution time, etc.

The drawback of this approach is the probable overhead that is anticipated during the execution of applications. in contrast, the tracing approach had a large number of overheads without impacting the regular execution of applications.

As can be observed, the online-based performance monitoring approach is advantageous for making early decisions. For instance, if an edge-AI application has a potential performance flaw, the entire application could be stopped at the initial stage of execution as the performance values will be visible to the user.

Figure 7.6 depicts the most commonly utilized performance monitoring approaches. The approaches are detailed in previous subsections.

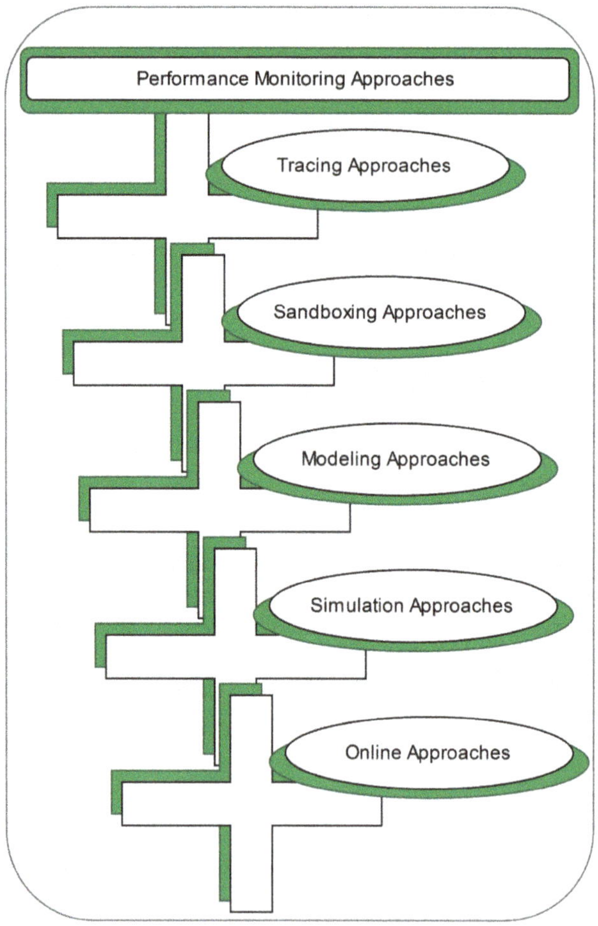

Figure 7.6. Performance monitoring approaches.

7.12 Cloud/fog/edge-level performance monitoring

Performance analysis tools are available as standalone devices that operate on clouds, fogs, or edge nodes. However, research efforts are in progress to develop an integrated performance analysis tool for an entire edge-AI application.

In the present situation, researchers apply available standalone performance analysis tools to monitor the performance of applications. For instance, tools such as AWS Xray, AWS cloud watch, Google Stackdriver, AppDynamics, Wavefront, Epsagon, Thundra, and Application Insights have manifested the performance monitoring of cloud-based applications. These tools are often utilized in computationally intensive clusters or fog nodes.

Additionally, tools such as Prometheus, InfluxDB, and Grafana have been utilized by several open-source enthusiasts in container-based working environments. However, there are very few performance analysis tools that are dedicated to the edge-AI computing continuum. The major drawback of designing such a performance analysis tool is that it scales the monitoring feature across multiple heterogeneous computing environments along with the applications. Not only that, the application has to perform the monitoring tasks without much overhead in real-time. Hence, the development of an end-to-end performance monitoring tool for edge-AI applications is still a research challenge.

7.13 Conclusion

Analyzing the performance of edge-AI applications can improve their applicability. Developers may need to focus on developing application logic rather than studying the impact of performance issues when executed on edge-enabled devices. This chapter studies the importance of performance analysis in edge-based frameworks. Additionally, it reveals the potential energy consumption issues of embedded devices and the techniques to monitor the associated issues. The chapter provides a taxonomy of performance monitoring in edge-AI domains.

References

[1] Ali-Eldin A, Wang B and Shenoy P 2021 The hidden cost of the edge: a performance comparison of edge and cloud latencies *Proceedings of the International Conference for High Performance Computing, Networking, Storage and Analysis* SC '21 (New York, NY: Association for Computing Machinery)

[2] Alves M P, Delicato F C, Santos I L and Pires P F 2020 LW-coedge: a lightweight virtualization model and collaboration process for edge computing *World Wide Web* **23** 1127–75

[3] Benedict S 2018 Prediction assisted runtime based energy tuning mechanism for HPC applications *Sustain. Comput. Inform. Syst.* **19** 43–51

[4] Benedict S 2023 Carbon neutrality approaches for IoT-enabled applications *2023 Third International Conference on Artificial Intelligence and Smart Energy (ICAIS)* pp 152–7

[5] Cassales G, Gomes H M, Bifet A, Pfahringer B and Senger H 2023 Balancing performance and energy consumption of bagging ensembles for the classification of data streams in edge computing *IEEE Trans. Netw. Serv. Manag.* **20** 3038–54

[6] Dirgantoro K P, Nwadiugwu W-P, Lee J M and Kim D-S 2020 Dual fieldbus industrial IoT networks using edge server architecture *Manuf. Lett.* **24** 108–12
[7] Guillén M A, Llanes A, Imbernón B, Martínez-España R, Bueno-Crespo A, Cano J-C and Cecilia J M 2021 Performance evaluation of edge-computing platforms for the prediction of low temperatures in agriculture using deep learning *J. Supercomput.* **77** 818–40
[8] Karzhaubayeva M, Amangeldi A and Park J-G 2023 CNN workloads characterization and integrated CPU–GPU DVFS governors on embedded systems *IEEE Embedded Syst. Lett.* **15** 202–5
[9] Kim J S, Javed K, Min K H and Roh J 2023 A 13.5-nA quiescent current LDO with adaptive ultra-low-power mode for low-power IoT applications *IEEE Trans. Circuits Syst. II: Express Briefs* **70** 3278–82
[10] Lange E O, Jose J M, Benedict S and Gerndt M 2023 Automated energy modeling framework for microcontroller-based edge computing nodes *Advanced Network Technologies and Intelligent Computing* ed I Woungang, S K Dhurandher, K K Pattanaik, A Verma and P Verma (Cham: Springer) pp 422–37
[11] Luan S, Hussain M, Ali S and Rahim S 2022 China's investment in energy industry to neutralize carbon emissions: evidence from provincial data *Environ. Sci. Pollut. Res.* **29** 39375–83
[12] Samikwa E, Di Maio A and Braun T 2022 ARES: Adaptive resource-aware split learning for internet of things *Comput. Netw.* **218** 109380
[13] Seetha R and Nandakumar R 2023 Benchmarking on RISC-V core and performance analysis of two open-source real-time operating systems *Proceedings of the NIELIT's International Conference on Communication, Electronics and Digital Technology* ed S N Singh, S Mahanta and Y J Singh (Singapore: Springer) pp 381–93
[14] Xia X, Chen F, He Q, Grundy J, Abdelrazek M and Jin H 2021 Online collaborative data caching in edge computing *IEEE Trans. Parallel Distrib. Syst.* **32** 281–94

Chapter 8

Security in edge-AI systems

This chapter investigates the security features and architectures of edge-AI applications or frameworks. In fact, security in edge-enabled computing systems is considered an essential component as data transports between edge and cloud data centers. Hackers are able to compromise edge devices due to the non-availability of sophisticated security-related algorithms. Hence, edge-level security components and their methods have to be investigated by edge-AI application developers or framework designers.

8.1 Existing security challenges

Edge-level security refers to security techniques that are practiced beyond traditional computing systems. This is because the security features have to be extended from cloud data centers to several remote edge nodes that are computationally incapable. The techniques that need to be developed have substantially evolved based on the existing security challenges in edge-enabled domains. The notable challenges are as follows:

1. Security attacks: In the edge-AI platforms, it is necessary to identify potential attacks such as DDOS, Sidechannel, etc, in real-time. However, there are notable challenges in quickly identifying attacks on edge-AI platforms owing to the ability to remotely access edge devices. The hackers concentrate on the inner details of devices and kill their operations. Similarly, it becomes difficult for edge nodes to monitor the security issues of these devices in real-time as they are geographically spread across multiple regions and fall under the jurisdiction of different organizations.
2. Data-specific issues: There is a challenge in attaining utmost data integrity, confidentiality, or prevention of data in edge-AI systems. The availability of data to process edge-AI applications is not an issue in many situations.
3. Multi-vendor issues: Edge-enabled devices belong to different vendors. The processing of instructions in Jetson Nano devices is different from processing

doi:10.1088/978-0-7503-5593-3ch8

instructions on Raspberry Pi-based devices. This shows that the security features of executing applications on these multi-vendor nodes can challenge the security components of edge-AI applications.

4. Heterogeneous issues: In the majority of cases, the devices have different processing components. For instance, there are GPU-based computing devices, CPU-based computing devices, or tensor units based on ASICs in edge-enabled frameworks. This means that the security requirements change across these heterogeneous devices.
5. Architectural issues: In several application domains such as agriculture, healthcare, finance, education, etc, there are not many generic edge-AI applications that suit well considering security issues or factors. The non-availability of generic architectures poses new challenges for architecture designers or security-related algorithm designers among the edge-AI application developer community.
6. Limited standards: The existence of standards to monitor and devise new security-enabled edge nodes is limited. Accordingly, there are several standalone solutions that cannot be compatible with the existing systems or devices. The device manufacturers are unable to design a generic device that fits in many edge-AI situations.

8.2 Security attacks in edge-AI

This section details the list of security attacks that hamper the growth of practicing edge-AI in various application domains. The security attacks are easily challenged by edge-enabled nodes. Hence, the characteristics of such attacks and the important remedies to solve them are detailed in this section.

Broadly, the security attacks in the edge-AI domain, where heterogeneous devices with varying computational capacities are involved, are classified into two categories:

1. Active attacks: Active attacks will have a direct impact on the security of platforms or devices. The attackers are keen on targeting or damaging the active system. For example, modifying an MQTT message by hacking the intermediate brokers of edge-enabled IoT systems is an active security attack.

   There are a variety of active security threats backing from the networking domain while executing edge-AI applications. Attacks such as distributed denial of service (DDoS), physical, side channel, malware injection, authentication/authorization, jamming, Trojan, routing information, forgery, and traffic injection. A detailed view of these active attacks and the procedures to avoid such attacks while executing Edge-AI applications is given in the following subsections.
2. Passive attacks: In passive attacks, the entire system that executes edge-AI applications will be passively monitored over a period of time. This could happen due to exposed ports or vulnerable cross-scripting issues in the networking platforms of the edge-AI framework. Passive attacks, in general, are harmful in future endeavors instead of having an immediate impact on the security of the systems.

8.2.1 Physical attacks

In physical attacks, devices are directly hacked or accessed by hackers. Also, they are physically attacked by security attackers. Notably, attacking a cloud server or edge-based device will hamper their normal functioning. In Raspberry Pi-based devices, the attackers could remove the SD cards or flash drives in a physical manner. These incidents could have a direct impact on the proper functioning of applications.

Similarly, sensors are often prone to attacks. For instance, a video processing-based edge-AI application will mount multiple cameras in risky locations. The attackers might physically target these camera sensors to stop future data transmissions. Also, near-field communication (NFC)-based devices are susceptible to theft as they are used for banking-related applications.

To avoid potential physical attacks, smaller, portable devices such as sensors and actuators need to be constantly monitored using external support measures. A dedicated team of security personnel may be deployed near to computers or servers. Indeed, the costs involved in physically protecting such devices could increase the budget involved in solving applications.

8.2.2 DDoS attacks

A distributed denial of service (DDoS) attack is one of the security vulnerabilities that disrupt the regular ongoing services of applications. In edge-AI applications, a large network of sensors, actuators, networked devices, and storage elements are involved. In a DDoS attack, an attacker tries to deliver unwanted services to the regular chain of work.

The DDoS attacks could be detected by evaluating the accessibility of edge-AI services or associated networks. The networks will be unexpectedly utilized by some other services hacked by a suspicious bot or personnel.

Detecting DDoS attacks in an edge-enabled application is a serious challenge. For instance, IoT-enabled services could have a surge in measurements at unpredictable times depending on the automated synergy of applications, especially in the healthcare domain. The delivery of services could be enacted by multi-vendors or multi-organizations. Hence, it is not a better choice to remove services due to the anticipation of DDoS attacks: a challenging task to identify such attacks. The authors of [13] predicted the potential DDoS attacks in edge-enabled applications by collaboratively involving multiple nodes and applying reinforcement learning approaches.

8.2.3 Side channel attack

A side channel attack aims at collecting valuable information from a system. In edge-AI systems, the side channel attack will collect all required details about sensor traffic or processing decisions. In embedded devices, it becomes easier for an attacker to capture the number of instructions executed by carefully watching the battery utility of systems or the power consumption of edge devices.

Depending on the gathered information, the attacker could disrupt the system in a timely manner. Or, the attacker could study the cryptoanalysis keys and distribution patterns elegantly.

A side channel attack could be carried out actively or passively. The most common practice is to manipulate the code sequences that influence the system. For instance, by evaluating the characteristics of the entire system, the attacker could modify the code by updating the over-the-air (OTA) updates.

A side channel attack is accomplished in several ways: (i) timing attack: this detects the modifications in the timing utilization, (ii) hardware monitoring attack: this detects the hardware utilization and the specific component utilization, and (iii) model sniffing attack: this investigates the model size, training time, and testing time of edge-AI applications for attacking the system.

In edge-enabled systems, the most common parameters that are utilized for side channel attacks are communication-related parameters, battery source-related parameters, and device monitoring information. Among these three parameters, the battery source-related parameters could have a strong influence on the attack designers owing to the weaknesses involved in such constrained devices. Attackers could flood hello messages or packets to inadvertently drain battery sources [9] to enable battery drains, which pave the way for such attacks.

In summary, the majority of side channel attackers are keen to observe the signals and signal to noise ratio (SNR) to capture the details. To override the situation, it is a better option to develop reconfigurable hardware that offloads processing power across multiple compute nodes. Also, it is possible to design algorithms that hop randomly from one machine to another to flee from the attacker's hold.

8.2.4 Malware-injection attack

Malware injection is a type of attack in which malware components are injected into edge-AI systems. The malware components, accordingly, hamper the devices by not providing the intended or useful services.

Malware injection is a serious threat to edge-enabled devices such as ESPs, where the code could be modified over the air. The concept of over-the-air (OTA) updates expects the delivery of updated code snippets from a few cloud-based services or edge-enabled services. An attacker could establish malicious code that serves as a source to eavesdrop on data or the learning model of edge-AI applications. The authors of [3] have listed over 64 malware attacks on IoT-enabled computing systems.

In distributed service-based computing environments, cross-scripting is a common malware attack. Here, Javascripts, VBScript, ActiveX, or similar scripts are maliciously injected into the active edge device. Accordingly, the scripts start to observe the characteristics of applications or intelligence performed in the system.

To avoid such malware injection attacks, machine learning-based malware detection systems have been developed in the past. ML algorithms such as support vector machines, random forests, Bayes algorithms, multi-layer perceptions, etc, have detected potential malware from machines by extracting the features and detecting anomalies due to the framed mathematical models. In some situations, instead of offering structured datasets to the machine learning models, some researchers have designed deep learning-based malware detection systems that could be deployed on edge-enabled systems to automatically capture the features and report the anomalies. As an advanced approach, a number of researchers have looked into offering an explainable AI feature to dictate the reason for classifying a service as malware. Although there have been a few research efforts in the past, malware injection attackers have a profound approach to tweaking the malicious services into edge-AI systems, considering the advancements of existing detection methods.

8.2.5 Authentication/authorization attack

In authentication/authorization attacks, an attacker aims to gain login credentials to access valuable resources. The authentication attack is, in particular, claiming that they are the valid user of the system. This could also be a bot engine in an edge-AI application that claims that the service is eligible to utilize the resources. In contrast, the authorization attack is the process by which users gain access to non-permitted resources.

The most common practice of any attacker of the authentication system is to utilize a specialized proxy-enabled service. This particular iterative service attempts to crack the user login details so that the login credentials can be utilized by a legitimate user. Hackers continue to monitor the user activities and develop a pseudo-login setup to collect the login credentials. In biometric-enabled systems, hackers parse the fingerprint details of a legitimate user from the biometric sensor to develop a similar kind of bio-fingerprint. Later, these fingerprints are used to access the entire application service.

In edge-enabled applications, the vulnerabilities due to authorization attacks are relatively higher than the other attacks owing to the lack of access control policies on smaller devices. The resources are prone to serious attacks, including VMrootkits, due to authorization attacks. To override VMrootkits, the authors of [6] have developed a lightweight virtual firewall approach that could continually trace the kernels for potential malware.

The most commonly seen reason for authorization attacks is due to broken access control mechanisms. Due to the challenged access control policies or approaches, attackers are able to remove resources from the vicinity of legitimate users. Additionally, the resources are claimed at higher costs in the future by the attackers.

The authorization-related attacks could be avoided by providing low privileges to users. In doing so, all users are not permitted to control or modify the resources necessary for executing an application. Another option is to avoid multiple attempts by the user while requesting resources. This prevents hackers from attempting to compromise the access control policies levied on a machine.

8.2.6 Jamming attack

Jamming attacks are a security issue in network-enabled applications or systems. Here, the devices are prevented from communicating with each other by inducing interference-based signals. The jamming attack is, in fact, a part of a denial of service (DoS) attack as the services are not reached by legitimate users. However, the difference between a jamming attack and a DDoS attack relies on the positioning of denials, i.e., the jamming attack targets the communication mediums of interconnected devices, including sensors or actuators.

The jamming attack is carried out in incremental steps. At first, the attacker reviews the networking credentials of any target network. Next, the attacker has to decide on the potential method of jamming the system. For instance, a jamming attack could involve random signals, constant pulses, or continuous evaluation-based signal floods. Next, the attacker has to choose an appropriate jamming device that can be hidden among the available edge systems. And, finally, the jamming process has to be accomplished for providing a total neglect of the denial of services.

To avoid jamming attacks, a few methods have been adopted in the past. They are listed as follows:

1. Channel hopping: In this technique, the intended channel frequency is hopped multiple times so that the attacker cannot jam the signals. The time required to identify the correct signal during a particular transmission will be so high that the attacker will not be in a position to jam them.
2. Jamming tracer: In this approach, any potential jamming-related incidents are tracked by specialized devices. The platform has to be vigilant in noticing newer signals that cause interference with the existing services. In fact, the tracing approach is a difficult task, and it is considered a logically complicated algorithm. The authors of [7] developed a jamming defense mechanism considering the Stackelberg Game theory.

8.2.7 Forgery attack

A forgery attack is a newly discovered attack that relates to the malicious delivery of system commands by trusted users of the system. The web services are attacked, and the associated resources are hampered by this attack. As an active measure, the attackers induce malicious data into the system and modify the learning models in the case of edge-AI applications.

A variant of hampering edge resources is found in modern web-related applications in which a request message is posted to users. When opening such websites, malicious code will be injected into the application to utilize the resources and damage the existing systems.

8.2.8 Passive security attacks

Active security-related attacks are harmful in several ways. However, passive security attacks can lead to a more harmful situation if they are not eliminated at the initial stage of the attack. A few notable passive security attacks and their anticipated vulnerabilities to edge-AI platforms are listed below:

1. Traffic analysis: This is a passive security attack in which the details of learning models or resource information are hacked by an attacker. The hacker monitors this information, which is similar to that of a normal, legitimate user. This camouflage of an attacker can become a harmful weapon in the later stages of deploying applications. For instance, the attacker can learn the traffic pattern or prediction pattern of a learning model; the bots can re-orient their reasoning by analyzing the existing reasoning patterns, etc.
2. Adversarial attacks: In edge-AI systems, adversaries are provided by sophisticated learning models such as generative adversarial networks. However, poor models or hacked machines could deliver the wrong adversaries to real-world applications. This method of providing wrong adversaries in a learning system due to an attacker is termed an adversarial attack. The adversarial attacks are often passive owing to the poor directions in a typical use case.

   The attacker could have full control or partial control over the learning models. Depending on the nature of the controls, the impact on the attack will be determined.
3. Eavesdropping or sniffing: An eavesdropping attack or sniffing attack is a passive security attack in which the data is parsed regularly over a period of time. The data, including a machine learning model, is passively parsed by an unauthorized user of the system. The parsing or sniffing happens at the juncture of communications between devices.

   The impacts of an eavesdropping attack could be avoided by utilizing a powerful device that can be dedicated to searching for such attackers. For instance, a firewall device or a device with apt-detection software could be solutions to avoid such attacks. In addition, edge devices could be prevented from accessing exposed ports or open services that are activated through mobility devices.

Figure 8.1 provides a pictorial list of the security attacks discussed in the previous subsections.

8.3 Data-specific vulnerabilities

Data-specific security vulnerabilities are security issues that relate to data associated with an application. In an edge-AI application, the security issues that are risky to the system due to data and their importance are described in the following subsections.

8.3.1 Data integrity-related issues

Edge-AI systems can fail due to data integrity-related security vulnerabilities. Usually, learning algorithms or training epochs generate models based on input data such as video frames, images, audio, etc. However, there is no strict policy to

Figure 8.1. Security attacks in edge-AI systems.

ensure the consistency of the learning outcomes. Due to the varying outputs in the majority of cases, the applicability of edge-AI systems is not a very secure solution.

In fact, data integrity has to be maintained while executing edge-AI applications, irrespective of the heterogeneous nature of systems or compute nodes.

To prevent security issues relating to data integrity, it is advisable to restrict access to fraudulent systems or bots. Also, it is a good solution to evaluate data and record the status of the data periodically. In this way, the potential issues could be avoided in an edge-AI system.

8.3.2 Data confidentiality-related issues

Data confidentiality is one of the crucial steps to prevent many potential security attacks in edge-AI applications. Data confidentiality is considered along with protecting files and resources in the system. It aims to prevent malicious users from eavesdropping on data. To ensure data confidentiality, encryption techniques are adopted between sender and receiver. The adoption of encryption techniques prevents unauthorized access to data by malicious users.

The necessity of providing data confidentiality in edge-enabled systems is to improve service availability and increase trust between users while utilizing the systems.

To improve the data confidentiality in edge-AI systems, the platforms should have mechanisms to limit access to authorized users; they should update security-related software or algorithms periodically; and they should support all possible security-related communication protocols so that many vendors can participate in edge-AI platforms.

8.3.3 Data availability-related issues

Data availability-related security issues pinpoint the prevention of accessibility for legitimate users. Hackers can attempt to intrude on the network and bypass the data availability for users. In some cases, delayed processing of data happens so that the data is not visible to the application, service, or user. In other situations, the data is hidden from the intended service paths.

There are a few notable reasons for the lack of data availability. For instance, frequent downtime of server/Docker instances hosted on computing machines, network-related issues that prevent accessibility of data sources, data misconfigurations that are not suitable over multiple networking platforms, etc.

8.3.4 Configuration mismatches

Edge-AI applications are prone to security attacks due to configurational failures. For instance, a microcontroller-based edge device may have to be configured considering the inner features of devices. However, those devices can be prone to several security issues if the configurations are not up-to-date. The authors of [11] noticed that a mismatch in configurations of edge nodes results in varying uplink rates and computational availability. Obviously, the resultant values are prone to unacceptable outcomes.

As edge-AI applications combine computing nodes from geographical locations, misconfigurations can lead to risky targets for potential hackers. This is a well-known problem in cloud-based distributed systems. For instance, modifying the networking policies of EC2 instances in AWS clouds can allow indefinite access to unauthorized users. Obviously, this can lead to potential security breaches in edge-AI solutions.

In some situations, the manual configuration settings might cause human errors. See figure 8.2 to view the data-specific security issues while executing edge-AI applications on platforms.

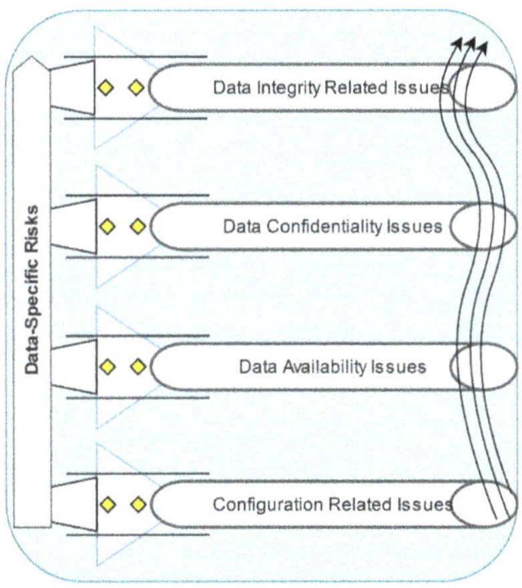

Figure 8.2. Data-specific security concerns.

8.4 Security architectures in edge-AI

Architectures are designed to efficiently deal with security issues and solve existing edge-level problems. In edge-AI systems, considering the security requirements of applications, security-enabled edge-AI architectures are designed by computer scientists and security-targeted enthusiasts [4]. Such architectures are crucial to designing performance- and energy-efficient solutions considering the changing future hardware or ecosystem. See figure 8.3 for a broad picture of the classification of security-oriented architectures in edge-AI systems.

In general, security-enabled edge-AI architectures could be classified based by the following factors:

- security-oriented functional layers in edge-AI,
- user perspectives,
- device-focused security aspects, and
- application-specific security strategies.

This section investigates these security-enabled edge-AI architectures and their involved entities in detail.

8.4.1 Security architecture: a layered perspective

Enabling security features in edge computing nodes can be accomplished by layering functional components such as controlling units, gateways, servers, etc. In a layered

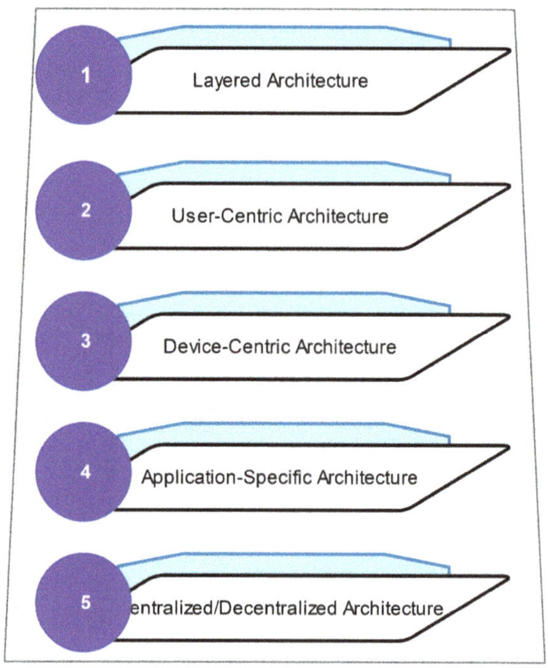

Figure 8.3. Security-enabled edge-AI architectures: types.

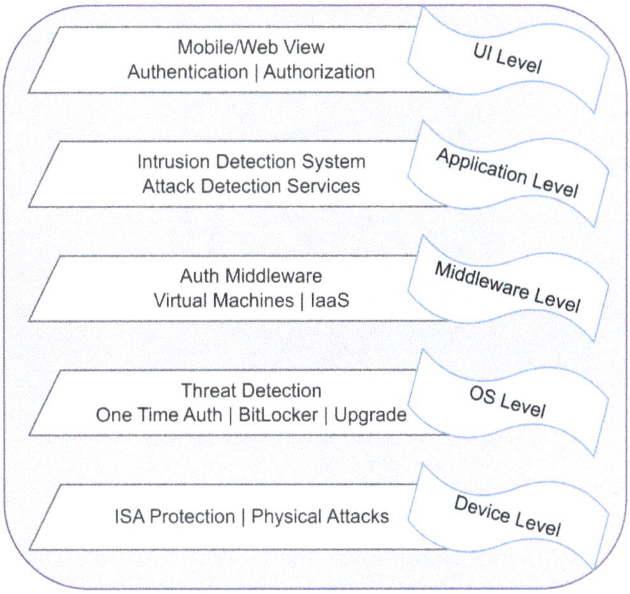

Figure 8.4. Layered security-enabled architecture for edge-AI.

architectural style, the functional modules are structured with a clear distinction between hierarchies or service deliveries.

A typical security-related layered architecture is illustrated in figure 8.4.

As shown in figure 8.4, functions relating to devices, such as device-level security features, are processed in the device layer. The OS controller units reside on top of the device-level security functions. The purpose of these controller units is to deliver smooth serviceability to the higher layers or device components. Notably, functions such as security violations, hardware-based random number generation, IP conversions, bitlockers, etc, are processed in layers. These services are distinguished from the other higher-level services levied by edge servers or cloud data centers. Edge servers will be responsible for automatic data transmissions, data leaks, authentication features, etc. Cloud data centers will provide long-term evaluations of security violations or prediction-related information. The user-interface-level security offers mobile and web views.

In short, it could be observed that the layered security architecture offers a distinct level of security-related services that can be scaled or modified in a modular fashion. An example of layered security architecture is discussed in [8]

8.4.2 Security architectures: user-centric

Designing security architectures in edge computing systems focused on users is another key aspect of improving security features. Users are interested in establishing all security features for their edge-AI applications. However, when budgets related to the convenience of utilizing edge devices or battery-operated components

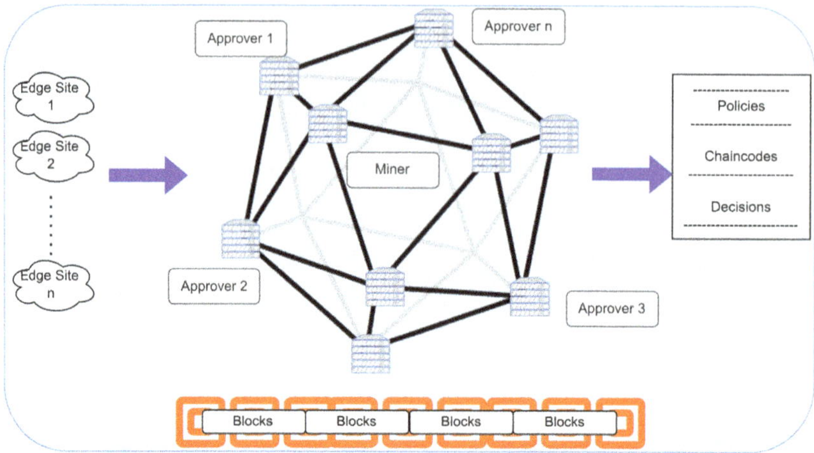

Figure 8.5. Blockchain-enabled trusted security architecture for edge-AI.

are considered, they attempt to minimize the security features. This is a known fact because increasing the algorithmic capabilities can drastically increase the costs involved in the design or architecture.

Users prefer to establish a trusted environment. The security-related architectures are expected to increase the trust involved in the systems. In traditional systems, the trust shared with users remained a challenging factor due to the increasing number of system failures or reliability issues. In modern user-centric security architectural designs, trust mechanisms are introduced using blockchains or similar immutable system designs. In blockchain-enabled trust systems, the entire trust mechanism is carried out by a group of networked nodes and user-levied policies. Obviously, such architectural designs have manifested ease and trustworthiness for the betterment and convenience of users.

Figure 8.5 illustrates the concept of applying blockchain networks that offer non-repudiated blocks of data using trusted blockchain networks.

8.4.3 Security architectures: device-centric

Security architectures have been developed with more emphasis on devices than focusing on users in a few edge-AI applications. The devices are often heterogeneous in nature and belong to multiple vendors without unique standards. Consequently, researchers have focused on developing security architectures for individual devices.

In these architectures, devices are bestowed with opportunities to modify the configurations and develop hardware-level protection systems. The devices need to adopt security features in their architectural design. If not, fundamental security functions can become an issue. For instance, a portable IoT-enabled key has to be prevented from connecting to unintended devices. To do so, the security features have to be powered at the hardware level. Also, if we consider the OTA updates of edge nodes, the devices have to be carefully allotted to utilize the memory space. If not, OTA updates could hamper the device facilities and induce malware into the system. Hence, all devices have to be securely designed at the hardware level to increase security features.

8.4.4 Security architectures: application-specific

Offering end-to-end security architecture for edge-AI applications is a normal expectation of users. This means that the applications could execute tasks securely without any limitations on enabling security functions. The security-related functions have to be distributed across multiple computing components, including end-users.

Typically, end-to-end security architecture in edge-AI systems includes a strong trust relationship in the underlying system. In order to incorporate a strong bond to the system, the architecture encapsulates identity management, attack protection schemes, device access policies, scalable security-related functions, and managerial functions. Also, the architectures involve at least a few dedicated security monitoring devices to learn the security-related characteristics of applications.

The identity management feature of this architecture will engage in detecting the authorized utility of bots or systems. It remains a crucial feature, as the solution could avoid unnecessary device fraud, phishing, and a few other security-related vulnerabilities. The architecture involves security-related functions to select an appropriate device for accessing policies and security-detection mechanisms. Ericcson [5] has explored the security architecture that was designed for applications in an end-to-end fashion.

8.4.5 Deployment-specific security architectures

Deploying resources are either centralized or decentralized. In centralized security architectures, the security-related functions of edge-AI systems are executed on centralized, powerful nodes. They are typically carried out on cloud-based, scalable computing nodes. The centralized security architecture focuses on a single administration to produce security-related decisions. Even if the security functions are developed as algorithms, they are processed and delivered under the purview of a single administration.

The advantages of having a single administration to deliver security-related functions are: (i) quick decisions, (ii) fast convergence of decisions, (iii) planned administrations or management, and (iv) uniform delivery of security-related services. However, the centralized approach to deploying security features can pose several security-related threats. For instance, the failure or security threat to one single administration, which handles the security advice, could lead to a complete disruption to the security system. This can hamper the entire application or device function.

Decentralized security architectures are often preferred by practitioners or device/function providers. In decentralized security architectures, multiple organizations are involved in security-related decisions. Similarly, distributed processing of security-related functions is possible. Hence, edge devices that are distributed across the globe to execute an application will be able to collaboratively make security-related decisions. Although there are several advantages to decentralized security options, there are several challenges that researchers have been working on in the past few years. They are listed as follows:

1. Poor Convergence: The security decisions will not converge as expected owing to the involvement of multiple administrative domains in the decisions. Hence, there is a possibility of delayed decisions while solving security attacks.
2. Varying Policies: The security-related policies keep varying over time and geographical regions. For instance, data or models developed from Indian datasets could not be shared with other countries due to privacy issues. Similarly, devices from China are not transported to many other countries due to security issues. All these varying policies delay the formation of security checks and decide on the related attacks.
3. Decision Duplication: The security-related decisions may be duplicated as the decentralized decisions are common in the decentralized security architectures. The decisions for the same security violation could be treated differently in multiple locations. This can lead to unfair treatment of decisions.

8.5 Preventing security breaches: strategies

Having observed several security attacks and types of security-related architectures, it is quite beneficial to understand the potential prevention steps and the methods that could be incorporated in the newer architectures. This section details the possible security avoidance strategies that could be adopted in edge-AI systems.

8.5.1 Protocol-specific assistance

There are a few data-specific high-level protocols to transfer data or learning models across networked IoT systems. For instance, notable protocols such as Constrained Application Protocol (CoAP), Extensible Messaging and Presence Protocol, Message Queue Telemetry Transport (MQTT), etc, have been developed to provide data services in edge-AI systems. Similarly, there are a few communication-specific protocols, such as long range wide area network (LoRaWAN), WiFi, Bluetooth LTE, 5G/6G, WirelessHART, etc, to connect with edge-based nodes.

In edge-AI systems, communication-related protocols could involve security-enabled protocols while enhancing the security components of applications. For instance, the Secured Message Queue Telemetry Transport (SMQTT) protocol or S-HTTP are examples of secured communication protocols in edge-AI systems.

8.5.1.1 SMQTT protocol

Usually, the MQTT protocol does not have features to secure data communications that are transferred between computing nodes. If applications require a strong security component, it is advised to include security features in the existing protocols, such as MQTT.

A few researchers attempted to solve the security issues of MQTT protocols by enabling VPN connections between sensors or actuators and brokers or between brokers and application services. Additionally, a few other researchers have transported MQTT messages over secure socket layers (SSLs) to increase the security features of applications.

In Mosquitto brokers [2], there is a provision to enable SSL/TSL configurations to provide security in MQTT messages. There are a few other brokers that are popular in MQTT frameworks: MuleSoft Platform, IBM MQTT, RabbitMQ, and meshIQ. All these MQTT brokers could be configured with SSL/TSL configurations to improve the security features of MQTT-based data transmissions in edge-enabled applications.

Additionally, S-MQTT has been developed as an extension to MQTT protocols in recent years. In S-MQTT, encryption algorithms are adopted in the form of lightweight implementations to accommodate the services on edge nodes. Appropriate encryption keys are transferred between brokers and publishers and subscribers using traditional key management systems. However, all these encryption algorithms are lightweight instantiations when compared to the traditional security functions adopted in traditional networks.

8.5.1.2 S-HTTP

Similar to traditional networks, S-HTTP is a protocol that offers secure hypertext over web services. Edge-enabled applications rarely follow a request–response model. However, there are applications and services that prefer to utilize the request–response model, as sensors or actuators may be powered using normal power grids. For instance, security cameras in airports or cautious locations will have to have a sufficient power supply. The packets could be transferred over WiFi systems using request-response models. However, to increase the security components involved in such applications, security-enabled protocols such as S-HTTP could be applied.

8.5.2 Reliable routing protocols

A few researchers have developed routing protocols that have multipaths to avoid security-related attacks. It is also applied to improve the energy efficiency of applications, so attackers could not harm the working system or devices. For instance, jamming attacks can harm the life of batteries. This is because the noise interference will be too high for the normal signals to be processed by any constrained devices.

Some other studies in the past characterized the security features of embedded devices by applying various routing algorithms.

8.5.3 Combination of lightweight and computationally-intensive implementations

Lightweight security algorithms are definitely the choice for edge-enabled systems. However, a pure lightweight security algorithm could not withstand all modern security-related attacks that are posed on edge devices. This is owing to the poor computational capabilities of edge devices. In the majority of cases, a combination of lightweight security-related algorithms and heavyweight computations in the cloud could be an appropriate choice.

In modern systems, lightweight block ciphers, lightweight stream ciphers, lightweight hashes, and lightweight cryptographic end-to-end solutions have been developed. The role of lightweight algorithms is to improve the memory space and computational space of algorithms. In fact, several researchers have put forth novelties in reducing the key sizes of the ciphers in a secured-edge-AI system.

The major objectives of establishing lightweight hashes are to minimize storage spaces and computing capacities. In a few implementations, a robust implementation of hashes is carried out in the cloud, whereas low-resource computations are practiced on edge nodes. In short, it is better to combine both lightweight and computationally-intensive algorithms with the assistance of an edge-to-cloud computing compendium.

8.5.4 Firmware-level updates

It is a known fact that firmware is software that has direct communication with hardware units. With the evolution of several edge-level computing nodes, firmware is increasing day by day. Typically, it is software that maps the requests of operating systems to underlying hardware.

In recent years, many embedded device developers have designed procedures to automatically update the firmware of sensor-enabled nodes. This feature was created in order to install these computational devices in unreachable locations. For instance, sensor-based devices that need to be mounted on machines that could hardly be reached by humans are a few examples. In such devices, the firmware needs to be modified for better functionality.

For instance, over-the-air update is a technique that is followed in several embedded-based edge devices to update the application code features of devices. It is also possible for hackers to update the firmware details during modifications.

It is, therefore, essential to prevent such wrong firmware updates. Researchers have previously used hardware-level procedures as well as network-level prevention mechanisms to accomplish this. The MAC addresses are mapped to specific authorized nodes to activate firmware updates. Otherwise, devices do not get regular updates. Thus, a few precautionary procedures have been practiced by developers in the past.

8.5.5 Prior security testing

Security testing is a common procedure to prevent security-related attacks in any network-related system, including edge-AI systems. The main purpose of having tests is to identify security risks, prevent near-future attacks, or remove potential hackers from the system.

Typically, security-related tests are practiced in several ways, as discussed below:

1. Application security testing: This testing is a step toward detecting potential security flaws in applications. Edge-AI applications have to utilize multiple

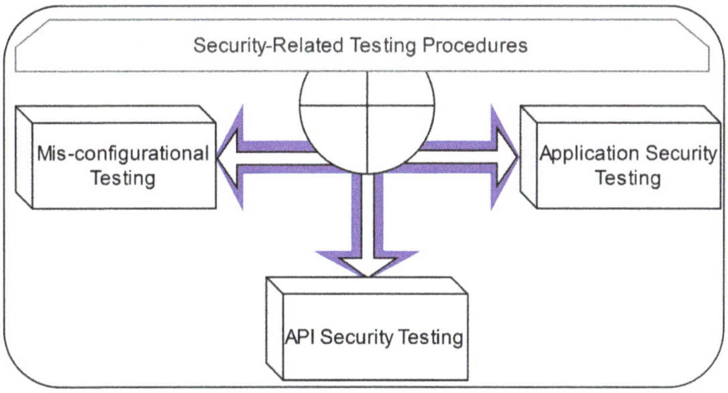

Figure 8.6. Testing approaches in security-conscious edge-AI.

computing devices. The software components, such as federated learning instances, have to surpass multiple nodes. There are several possibilities for potential security threats. Hence, application security testing techniques could look into the real reason for security flaws, if any, in the system.

2. API security testing: This testing approach handles the issues relating to associated application programming interfaces (APIs). The majority of the edge-AI applications are not standalone applications. They will require several third-party library services. Usually, such APIs are prone to several attacks, such as man-in-the-middle attacks. The security challenges posed by such APIs need to be carefully evaluated and identified by testing procedures.
3. Mis-configurational testing: In distributed systems, especially in embedded devices, there are many locations to either manually configured or automatically configured systems. For instance, setting up sleep modes, managing memory spaces, etc, are configured using some configurational methods. There is a higher risk of wrongly configuring such files. The potential of human errors could lead to intolerable security vulnerabilities. Hence, the mis-configurational testing approach attempts to look into potential configurational issues in devices.

The pictorial representation of these three security-based testing components of edge-AI systems is given in figure 8.6.

8.5.6 Encryption approaches

Obviously, it is a clear indication that traditional cryptographic solutions need to be incorporated in edge-enabled devices. However, to meet device resource constraints and battery limitations, the encryption techniques have to be lightweight and robust in all aspects.

There are two approaches to developing cryptographic solutions: the hardware approach or the software approach. In hardware approaches, encryption techniques are developed such that newer components are introduced to process encryptions.

For instance, to develop a stream cipher, specific hardware registers could be designed with filters to map input bits while developing the crypto outputs. Similarly, FPGA-based hardware accelerators could be employed to speed up the logistic mapping processes while developing cipher outputs.

Apart from hardware approaches, there are several algorithmic approaches that follow encryption techniques. For instance, authors have developed an algorithm that modifies the keys regularly in order to protect against key-related attacks. There are also efforts to protect keys, considering memory and battery utilization. Another method is to apply the traditional substitution ciphers in blocks, considering the battery utilization of devices. In a few works, researchers have developed no-pairing encryptions based on the elliptic curve decision-based Diffie–Hellman approach.

8.5.7 Newer technologies

There are a few methods to apply novel technologies to protect keys and input data in edge intelligent systems. The most widely applied technique is establishing blockchains to improve the trust involved in systems. However, deploying blockchain nodes across multiple edge devices is a challenging problem. For instance, which edge node can execute blockchain software components such that the blockchain ledgers are not hampered by attackers? This type of question is studied in newer security-enabled IoT architectures.

The blockchain technologies ensure data protection while transmitting models across compute nodes. For instance, in hyperledger-based blockchains, each transaction is recorded in the ledger records after being verified and approved by a group of approvers. This approach is applicable in edge-AI distributed systems, especially when data models need to be transported across multiple edge nodes. If the size of the model is higher for recording the transactions in blockchains, it is possible to include an inter-planetary file system approach where hashes of the entire blocks are recorded in blockchain ledgers. The real properties of the hashes will be stored in remote locations or cloud systems for utilization.

8.6 Tools and solutions

To adopt security-based systems, it is useful if the mechanisms are readily available for users in the form of production-level tools. The security tools could be developed as libraries, plugins, standalone applications, and hardware devices. In this section, we shall learn a few readily available security-enabling tools for edge-AI systems.

8.6.1 FireEye

FireEye is a cloud-based security tool that is effectively utilized for edge-enabled applications. FireEye is a company that offers security solutions for industries, with more emphasis on the industrial IoT sector. It has several automation features that could be easily combined with industrial sensor equipment to impart security features. FireEye-based security solutions are well recognized owing to their flexibility in combining intelligence at edge-level computations. The authors of [12] have applied FireEye to an unmanned aerial vehicle (UAE) delivery system.

8.6.2 PaloAlto networks

PaloAlto Networks Inc. [1] has developed several cloud-based security solutions that could be utilized for IoT-based edge devices. One of the crucial solutions that PaloAlto offers is the real-time inventory of IoT-enabled edge devices through their firewall security features. Their solutions are developed as APIs. Therefore, edge-based applications could integrate their services with the applications by invoking specific APIs. The APIs have functions to cross-check the validity of edge devices while utilizing the associated networks. The tool has the capability to check the risk involved in connecting an edge device by specifying the `risk-score` API component.

The major attraction of this tool is its capability to collect the MAC addresses of connected sensor nodes or edge devices within its purview.

8.6.3 Symantec solutions

Symantec Inc. has integrated security components into IoT-assisted edge nodes in cloud systems. The important features of its developments are to develop endpoints that prevent network or USB-related security attacks. Their major focus is on industries and allied sensor-enabled devices. The security-based solutions of Symantec could be applied to incorporate security features into edge intelligence in industrial IoT sectors [10]. For instance, security solutions have been mounted on various industrial sectors such as the cloud manufacturing sector, the oil and gas sector, the drilling sector, the healthcare sector, etc. The industrial control system protection based on the USB scanning approach has attracted several researchers and practitioners owing to its ability to protect the attached storage units.

8.6.4 Security: federated systems

Apart from the tools that are specific to edge-enabled systems, there have been efforts to design security features for federated systems. Edge-AI systems have utilized federated engines to increase the accuracy of learning models by utilizing local datasets. Hence, security-based solutions that focus on federated systems have become an important ingredient in modern applications. The most notable security-based solutions that are specific to federated edge learning systems are listed below:

1. BlockFL: Blockchained ondevice federated learning is an approach to solving the problem that evolves due to untrustworthy devices. This security solution is an academic one that attempts to apply blockchain technologies to secure the privacy of data.
2. FedCS: Federated learning with client selection (FedCS) is an approach that chooses edge devices based on security choices: i.e., it selects devices only if they are suitable to prior policies that are designed by the users. In this way, authentication and authorization of devices are inherently adopted by applying the FedCS solution.

3. FedPer: Federated learning with personalized layers (FedPer) is a technique to restrict the learning layers of deep learning models from public access. The major idea behind developing such a solution is to split the deep learning layers so that a few learning layers are prevented from unauthorized access, i.e., these personalized layers are subjected to utilizing only the datasets that are approved by users. Hence, the personalized layers are allowed to localize datasets while learning the models.

8.7 Conclusion

Edge-AI systems are prone to potential security issues that are difficult to detect and rectify. This is due to the fact that there is not enough space and computational capability on edge devices when compared to clouds. This chapter discusses the potential security attacks that hamper edge-AI applications. Additionally, the chapter provides a taxonomy of security-related edge architectures that have been studied by researchers in the past. Readers gain considerable insight into security features practiced in edge-AI systems.

References

[1] Alto P 2024 IoT—docs.paloaltonetworks.com (https://docs.paloaltonetworks.com/iot) (accessed 31 January 2024)

[2] Broker M 2024 Documentation—mosquitto.org (https://mosquitto.org/documentation/) (accessed 20 February 2024)

[3] Gulatas I, Kilinc H H, Zaim A H and Aydin M A 2023 Malware threat on Edge/Fog computing environments from internet of things devices perspective *IEEE Access* **11** 33584–606

[4] HaddadPajouh H, Khayami R, Choo K-K R and Parizi R M 2020 AI4SAFE-IoT: an AI-powered secure architecture for edge layer of internet of things *Neural Comput. Appl.* **32** 16119–33

[5] Suihko T, Mononen K and Teppo P 2017 Ericcson's end-to-end security management for IoT (https://www.ericsson.com/en/reports-and-papers/ericsson-technology-review/articles/end-to-end-security-management-for-the-iot) (accessed 31 January 2024)

[6] Li Y-G, Chung Y-C, Hwang K and Li Y-J 2021 Virtual wall: filtering rootkit attacks to protect linux kernel functions *IEEE Trans. Comput.* **70** 1640–53

[7] Liu J, Wang X, Shen S, Fang Z, Yu S, Yue G and Li M 2022 Intelligent jamming defense using DNN Stackelberg game in sensor edge cloud *IEEE Internet Things J.* **9** 4356–70

[8] Qiu T, Chi J, Zhou X, Ning Z, Atiquzzaman M and Wu D O 2020 Edge computing in industrial internet of things: architecture, advances and challenges *IEEE Commun. Surv. Tutorials* **22** 2462–88

[9] Ioulianou P P, Vassilakis V G, Smith R, Palin D and Shahandashti S F 2020 Battery draining attacks against edge computing nodes in IoT networks *Cyber-Phys. Syst.* **6** 96–116

[10] Symantec 2024 Hardening internet of things and operational technology (https://docs.broadcom.com/doc/iot-data-sheet) (accessed 20 February 2024)

[11] Wu H, Tian H, Nie G and Zhao P 2020 Wireless powered mobile edge computing for industrial internet of things systems *IEEE Access* **8** 101539–49

[12] Yao A, Jiang F, Li X, Dong C, Xu J, Xu Y, Li G and Liu X 2021 A novel security framework for edge computing based UAV delivery system *2021 IEEE 20th International Conference on Trust, Security and Privacy in Computing and Communications (TrustCom)* pp 1031–8

[13] Zhou H, Zheng Y, Jia X and Shu J 2023 Collaborative prediction and detection of DDOS attacks in edge computing: a deep learning-based approach with distributed SDN *Comput. Netw.* **225** 109642

Part III

Tools and solutions

Chapter 9

Frameworks: edge-AI platforms

Edge-AI frameworks, in contrast to the edge-AI architectures, relate to the existing edge-AI solutions that are available for utility purposes. These frameworks have a few pre-built functions or libraries that offer peculiar capabilities while extending the AI features of edge-enabled applications. This chapter explains the most commonly available frameworks that can be tailored to specific applications.

9.1 Essential characteristics

To develop edge-AI applications in various sectors such as agriculture, healthcare, and others with ease, edge-AI frameworks have been developed by a few practitioners and researchers. The frameworks in edge-AI applications refer to partial implementations that could be further developed or tailored to specific problems.

The notable essential characteristics of edge-AI frameworks in developing edge-AI applications are listed below, along with the details (see figure 9.1):

1. Autonomous: In general, autonomy is a feature that allows independence in operating edge-AI functions and performing decisions without the influence of associated systems. The granularity of implementing autonomous decisions is one of the key research aspects of edge-AI systems. For instance, deciding on the adversaries of time-series datasets could be at various levels. These levels of adversaries are variably performed in learning models such as Hive-Cotev2, InceptionTime, and Captum-python explainable AIs. The edge intelligence derived from learning models or adversaries has to be prevented from potential adversarial deadlocks.
2. Real-time: The framework should have the capability to develop and deploy learning models or federated solutions in real-time. The requirements have been bolstered due to the evolution of modern applications that utilize varying computational embedded devices. One of the reasons behind adopting real-time edge-AI frameworks is to engage dynamic developers and their solutions without much variation to any real-world use-cases. In

doi:10.1088/978-0-7503-5593-3ch9

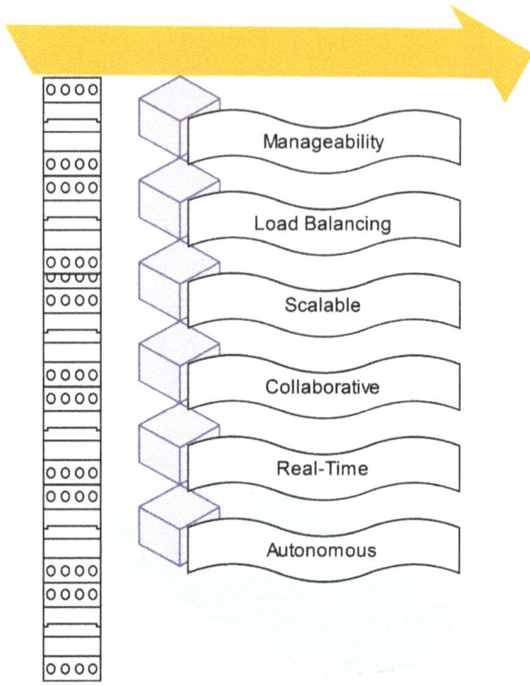

Figure 9.1. Characteristics of edge-AI frameworks.

fact, there are several edge-AI applications that require real-time processing capability [2, 12].

3. Collaborative: In the modern world, teamwork is more important for constructive growth than being a single performer. Similarly, edge-AI applications have to deliver functions in a collaborative manner. For instance, federated learning models will utilize multiple computing machines and distributed datasets to develop a robust learning model.
4. Scalable: Scalability is one of the parameters of AI solutions. The frameworks have to scale tens of thousands of edge devices or sensor networks while gaining learning insights. The value of such frameworks will reap many benefits for training models or inferences. The frameworks have to produce the same quality of results in terms of prediction accuracy or learning rate, even if they are scaled to multiple machines or algorithmic progressions.
5. Load balancing: The resources are widely spread across a large geographical location. Hence, it is important to balance the workloads or tasks among the available resources. In general, load-balancing-related algorithms may be developed to continuously monitor the availability of resources and their workload before assigning tasks to them.
6. Manageability: In some cases, the manageability of learning models or devices can be an arduous task while scaling frameworks. Hence, a framework has to consider the manageability of porting algorithms or allocating resources while executing edge-AI applications.

9.2 Types of framework

In the past, frameworks have been designed for achieving most of the essential characteristics specified in section 9.1. Depending on the execution pattern, resource delivery mechanism, learning methods, and so forth, the available edge-AI frameworks are classified into four broad categories, as follows:

1. Resource-allocation frameworks: The frameworks focus on allocating resources to edge-AI implementations. For instance, the frameworks can improve the energy efficiency or battery utilization of resources by distributing tasks to appropriate compute resources. In this case, the emphasis of designing resource allocation frameworks is on improving resource efficiencies.
2. Cloud-specific frameworks: There are frameworks developed by cloud providers such as Amazon, Google, Microsoft, Cisco, VMWare, etc. The major benefits of these cloud providers are that they extend their cloud capabilities to edge nodes while imparting AI or deep learning algorithms. The providers have the inherent objectives of increasing the profits involved in executing edge-AI applications and accommodating multiple users by efficiently handling resources.
3. Application-specific frameworks: A few frameworks focus on improving the performance of edge-AI applications. Notably, they are designed specifically for various application domains such as agriculture, industry, education, finance, etc. Accordingly, the frameworks will have specialized software or hardware components that increase the functional delivery of resources to tasks.
4. Distributed federated learning frameworks: A large sector of researcher has been engaged in developing federated learning-based edge-AI applications. Their emphasis is on increasing the utility of multiple edge nodes to create learning models and utilize local datasets with combined learning models.

The following sections will discuss the most commonly available frameworks that could be utilized in the above-mentioned categories. In addition, the probable challenges faced by researchers or practitioners in these categories will be described.

9.3 Resource-allocation frameworks

Resource-allocation Frameworks in edge-AI are solutions that enable distributed sharing of resources while executing edge-AI applications. The frameworks deal with the large-scale deployment of AI models or layers of deep learning models, considering the computing requirements of tasks.

In deep learning algorithms, layers can be decomposed into multiple components so that the layers are distributed to multiple computational devices. A number of researchers have named this decomposition of deep learning layers as split computing. In general, all deep learning algorithms, such as convolutional neural networks, reinforcement learning algorithm, recurrent neural network (RNN), fast-CNN, etc, have many hidden layers to calculate weights and classify images. With the concept of splitting layers and allocating these tasks to different computing devices, the end-to-end AI application will have opportunities to execute them with higher performance and privacy.

For instance, the SplitDDL video algorithm was developed by researchers to distribute hidden layers of deep learning algorithms to edge-enabled devices. They studied the performance of learning models and the other resource-allocation-related performance metrics of distributing edge-AI workloads [22].

While executing decomposed deep learning layers, it is possible to distribute the workloads to existing inexpensive computing devices such as ESPs or Raspberry Pis instead of delivering them to expensive nodes such as Jetson Nanos. The resource allocation frameworks will have libraries or pre-built APIs to share resources or services. Also, frameworks can select renewable energy-powered devices while developing learning models.

Typically, frameworks have specific schedulers to distribute such deep learning layers to multiple computing nodes. Additionally, edge-AI tasks such as sharing learning models, updating learning models, upgrading packages, etc, require distributed resources to be shared. These tasks need not be a portion of the hidden layers of deep learning. Instead, the tasks could be other deep learning algorithmic tasks or model development-related tasks.

Usually, in edge-AI systems, the resources are spread across distributed regions. Balancing the edge resources, considering the heterogeneous features of hardware units, is important in such environments. This can significantly improve the overall performance of edge-AI frameworks. The best practices, such as adopting a good load balancer, replicator, or caching mechanisms, could improve the performance of such edge-AI frameworks.

The allocation of resources in edge-AI frameworks could become a challenge for the following reasons:

1. Heterogeneity: Resources utilized in edge-AI systems are highly heterogeneous in nature. The resources have varying CPUs, memory units, architectural hierarchies, storage support, etc. For instance, ESP32 has Xtensa processors; Raspberry Pi has ARM-based processors; and Intel Movidius has Intel processors. These heterogeneous devices are utilized in single-distributed edge-AI applications. Obviously, they can create complexities in frameworks while assigning compute resources.
2. Non-availability: There is a high dynamism among the heterogeneous systems of edge-AI frameworks. The devices could be powered ON or OFF owing to the unreliable networks or battery-operated devices. Accordingly, the frameworks will have to adopt novel techniques to counteract the non-availability issues.
3. Unpredictability: The resources are highly unpredictable for the framework. The characteristics of resources could change depending on several factors. For instance, resource capabilities vary based on battery availability,

computational power, or hardware limits. Due to the high level of unpredictability, the edge-AI frameworks will have to have a robust design that can handle such situations.

4. Configurational failures and complexities: Each edge node is prone to various failures or non-availability issues owing to misconfigurations. The failures in the configurational settings are due to manual setup or automatic engines that execute configurational setups. Obviously, preventing such failures will require a complex design for edge-AI frameworks.
5. Privacy and security failures: There are possibilities in edge-AI frameworks to encounter data leakages as many tiny devices are involved in the learning processes. The models might be challenged by involving peer networks in some cases. For instance, a federated learning environment will have to update the learning inferences or weights on cloud devices in the majority of cases. An attacker could keep track of open ports in cloud services while updating the data models. A detailed description of the security challenge is studied in chapter 8.

Figure 9.2 highlights the challenges faced by edge-AI systems while allocating resources for executing them.

The resource allocation frameworks are subdivided into various subcategories, such as: (i) real-time allocation framework, (ii) cooperative resource allocation framework, (iii) collaborative resource allocation framework, (iv) QoS-aware resource allocation framework, (v) optimal resource allocation framework, and (vi) end-to-end resource allocation framework. The details of these resource allocation frameworks are discussed in the following subsections.

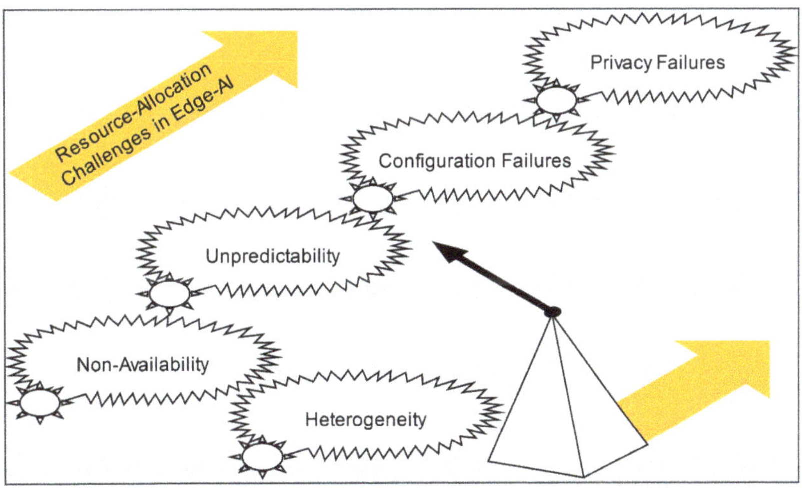

Figure 9.2. Resource allocation challenges in edge-AI.

9.3.1 Real-time resource allocation framework

The resource allocation in edge-AI systems has to involve multiple parameters such as heterogeneity, QoS, privacy, unpredictability, etc. One of the parameters that is crucial in edge-AI systems is the ability to select appropriate resources in real-time. To do so, a few software components have to be incorporated into the framework. Along with their functionalities they are:

- Handling real-time data: The framework has to handle real-time data from local systems: i.e., the data from sensors or similar sources has to be immediately available to training models or federated learning engines. The crucial importance of having such processing elements in the framework is that they can deliver time critical applications such as weather forecasting or camera surveillance-related applications. Hence, data analytics can be handled in real-time using such frameworks.
- Handling privacy issues: The resource allocation framework should have components to deal with privacy-related issues in real-time so that the anticipated attacks can be minimized. To do so, functions such as identification, traceability, authorization/authentication, and data manipulation identification have to be incorporated into the framework.
- Reduced resource outages: There is a rise in situations where resources could go out of service. In a real-time resource allocation framework, mechanisms need to be developed to handle such resource outages or identify the root cause of problems in real-time.

A typical real-time resource allocation framework collects sensor data from sensors and processes data nearer to the sources. This enables the privacy and protection of local data. The processed data is transferred to clouds using communication protocols such as MQTT. The services are handled in a publish-and-subscribe model. The clouds have connectivity with several edge-level compute nodes. Hence, they have to provide a scalable environment with robust load balancing algorithms or learning features. The pictorial representation of the real-time resource allocation framework is shown in figure 9.3.

Some of the identified challenges of implementing such a real-time resource allocation framework for edge-AI systems are: (i) predictability, (ii) maintaining

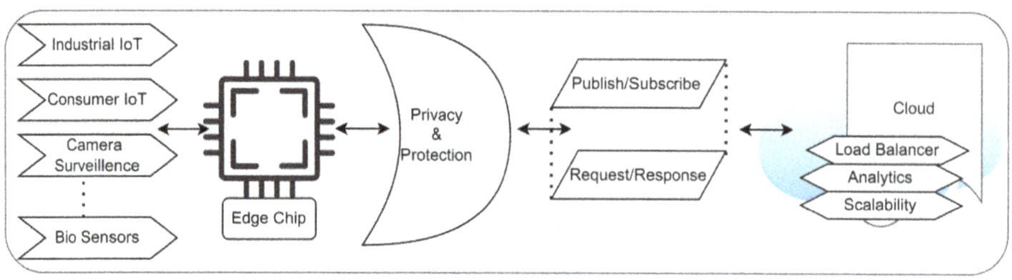

Figure 9.3. Real-time resource allocation framework.

timing constraints for tasks, (iii) allotting appropriate resources by scheduling tasks, (iv) identifying available resources, and (v) ensuring robust security features.

9.3.2 Cooperative resource allocation framework

In a cooperative resource allocation strategy, resources involved in edge-AI decisions are allocated based on the collective behavior of a group of resources or resource providers. In edge-enabled systems, there are resources from multiple parties or vendors to execute tasks. In traditional methods, a single algorithm decides on delivering resources to accomplish decisions in edge-AI applications. However, in the cooperative resource allocation approach, multiple providers are involved in deciding the utility of resources.

Often, the cooperative resource allocation framework employs a game theory mechanism in deciding on resource allocation decisions. In particular, cooperative game theory is a concept in which multiple players jointly take decisions to win over the opposing player. Although collective minds are involved in the cooperative game theory approach, the solution has to equally share the winning criterion. For instance, in edge-AI frameworks, the cooperative resource allocation framework that applies the cooperative game theory concept will have to reserve successes for the joint efforts of involved organizational domains.

The cooperation of resources for solving edge-AI-related problems is required when multiple resource providers are involved in the framework. For instance, assume a situation that involves an AWS-based Greengrass solution and Google's metaverse services to predict the identified objects of a picture. There is a requirement for sharing resources between multiple providers to execute the application. However, the utility of resources has to be fairly shared between the providers if the underlying application will reap monetary benefits.

The application of cooperative resource allocation in edge-AI applications is quite common when mobile networks are involved in sharing resources. Notably, mobile network-related infrastructural resources belonging to different network providers, such as BSNL, Telekon, Vodophone, etc, could not be laid out in all geographical regions across the globe. The resources will have to be cooperatively distributed among all providers involved in executing edge-AI applications.

In some instances, researchers have developed a hierarchical level of cooperation while allocating resources to solving edge-AI applications. The hierarchies can impart administrative freedom at each level of the framework. At each level of resource allocation systems, the number of resources that need to be delivered or profit sharing procedures and policies could be varied to increase the scalability feature.

Some of the distinct components that are required in realizing the development of a cooperative resource allocation framework are: (i) agents that could detect the number of cooperative resources, (ii) policies or procedures for distributing cooperative resources that belong to multiple administrative domains, and (iii) value-sharing mechanisms among the group of cooperative resources that engage in the edge-AI application.

As the cooperative resource allocation framework has to deliver cooperative resources, identifying and registering them in a common pool of resources can be a challenge. Additionally, there is a certain level of complexity in the framework if all agreeing parties cannot confirm the return values due to cooperation.

9.3.3 Collaborative resource allocation framework

The collaborative resource allocation framework focuses on developing learning models in association with different levels of computing systems. The learning models are developed in a collaborative manner by transferring the models developed by the framework to different nodes. The computing nodes could be the nodes from edge networks, fog networks, or cloud datacenters. The main focus of this framework is to allocate resources effectively so that the learning models are completed in a short span of time [21].

9.3.4 QoS-aware resource allocation framework

The QoS-aware resource allocation framework attempts to iteratively improve the service delivered by frameworks. The users of edge-AI applications have different viewpoints on delivering resources. However, the users aim to increasing the performance of applications for a given volume of resources.

The authors of [4] have developed a deep reinforcement learning-based resource allocation method named 'DeepEdge' by maximizing the interests of users. The primordial objective of the DeepEdge framework is to monitor the available resources for different edge-enabled applications and allocate appropriate resources. In the framework, mapping components are tested to effectively map the requirements of edge-enabled tasks and the availability of resources from various locations.

The utilization of the mapping component in the QoS-aware framework can be fine-tuned based on various performance metrics such as utilization of resources, availability of appropriate resources, energy utilization of resources, data affinity of compute machines, etc.

However, the framework has a few potential drawbacks. The development of an efficient mapping system is a challenge. This is because the clock involved in collecting the requirements of tasks or resources has to be synchronized properly before mapping them. Also, collecting a clear picture of available resources in real-time and capturing the requirements of resources from various heterogeneous edge-AI tasks could be considered challenges.

9.3.5 Optimal resource allocation framework

Optimization in a resource allocation framework refers to distributing resources considering several conflicting objectives [15]. The objectives are minimizing the energy consumption of applications, maximizing resource utilization, minimizing the execution time of applications, etc. In many applications, there are tradeoffs in the underlying optimization objectives. For instance, minimizing the energy consumption of applications could attempt to increase their execution time. Also, in some situations, the utilization of resources increases the energy consumption of applications.

Hence, optimal resource allocation frameworks designed for edge-AI applications have to undergo the following conditions:

1. Dynamic decision-making: Before formulating newer policies, it is better to decide whether the resource allocation framework has to follow dynamic or static conditions. In fact, many edge-AI applications are inclined to the dynamic nature of sharing resources. Hence, the framework has to have methods and techniques to deal with the dynamic situation of sharing resources.
2. Alternative matching resources: The optimal allocation of resources will require sufficient alternative resources to attain the objectives. For instance, Docker containers may be replicated on different servers to increase the redundancy of matching resources for edge-AI applications.
3. Transparent allocation: The number of choices can grow tremendously in a highly dynamic environment of edge-AI. Hence, it would be necessary for the frameworks to highlight the reason for choosing an appropriate resource.
4. Termination criterion: The frameworks will require mechanisms to terminate the selection choices. If not, the selection algorithms could continually decide on the best possible resources for edge-AI applications without proper convergence. The termination criterion of such frameworks could be decided by monitoring the non-improvement in the selected objectives.

9.3.6 End-to-end resource allocation framework

The end-to-end resource allocation framework aims at developing methods and procedures that offer a comprehensive solution. The solution that considers all details at different levels and stakeholders is termed an end-to-end framework. The users, computing systems, clients, analytic engines, communication systems, and monitoring units are different entities that are involved in the end-to-end frameworks.

The purpose of designing an end-to-end framework is to assess the comprehensive values of the framework while executing applications. In an end-to-end resource allocation framework, the costs involved in allocating resources for edge-AI applications could be evaluated using pre-built libraries or APIs. Similarly, all other performance metrics, such as energy, hardware cache misses, execution time, etc, considering all involved services or hardware components, could be assessed using the end-to-end evaluation framework.

Google has focused on developing an end-to-end edge-AI framework where model training and testing are assisted with sophisticated AI tools and Google cloud platforms The model training and testing phases on Google platforms are scaled using Kubernetes and Docker container instances.

Figure 9.4 provides an bird's-eye perspective on classifying the edge-AI frameworks that focus on resource allocation strategies.

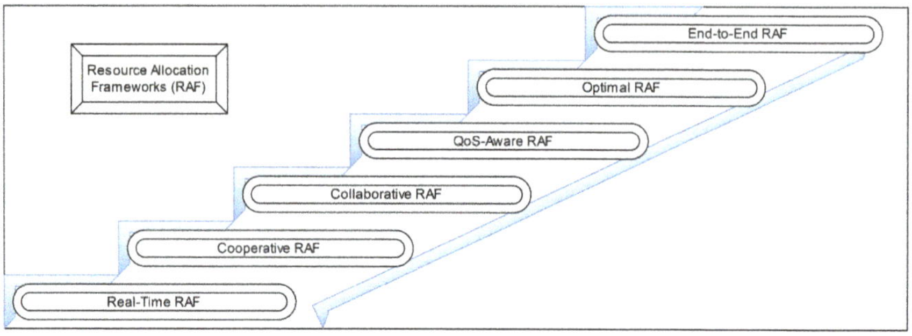

Figure 9.4. Bird's eye view for classifying edge-AI frameworks based on resource allocations.

9.4 Cloud-specific frameworks

Edge-AI algorithms have been widely practiced by cloud developers in the recent past. Notably, cloud providers, namely Microsoft, Google, Amazon, and others, have extended their services to support edge-enabled applications. Traditional cloud services such as Infrastructure-as-a-Service (IaaS), Platform-as-a-Service (PaaS), and Software-as-a-Service (SaaS) have been moved to edge nodes for enabling edge-based solutions. In this section, we will learn about the potential developments or products that are extended from the cloud to the edge by potential cloud providers.

9.4.1 Microsoft Azure-based frameworks

Microsoft has developed cloud-based solutions using Azure products. Infrastructure-related services such as establishing virtual instances of machines, storage instances, and dynamic networking services have attracted a large number of researchers and practitioners to work on cloud-based solutions. All these traditional cloud-based services expect the users to utilize compute or storage resources that are hosted in Azure-controlled third-party environments.

With the modern requirement of edge-enabled systems, Microsoft has extended their solutions to edge nodes: i.e., services are hosted nearer to the data on the user's computing infrastructure [8]. More specific to edge-AI solutions, Microsoft has developed a product named 'Azure Stack Edge'. This product is a specialized device with hosted services to diligently transfer limited data to cloud. Many data preprocessing steps are handled by this device. In doing so, the network bandwidth is improved and the performance of inferencing at the edge is enhanced.

The Azure Stack Edge product has a few services hosted on Docker containers to deliver data models to Azure IoT hubs. The most important functions of the developed product are to pre-process data and transfer it to Azure IoT hubs that are hosted on clouds. The advantages of these cloud-based solutions are the amalgamation of different cloud-based services, such as identity services or authentication services, along with edge-AI applications.

Microsoft Azure has offered a few other extended services using the Azure Stack Edge device [3]. Notably, services such as offline processing and uploading, synchronizing data, and supporting AI accelerations are considered vital for the product. Recently, Microsoft has developed a Stack Edge Pro 2 solution that combines the Stack Edge device with other cloud-related tools. The support for edge-AI using Azure-related products is served geographically by splitting the regions into the East US, West EU, and South East Asia regions. The regions have been split in order to deliver high-performance services considering the space, resource, and administration availability of systems.

9.4.2 Google-based edge-AI frameworks

Google has developed an edge-AI framework along with a few specialized hardware devices. It has developed the Coral platform to enable edge-level inferences based on local data [1]. The idea behind developing the Coral platform is to preserve the privacy of edge-enabled devices. The framework involves hardware components such as edge accelerators, software components such as AI algorithms, and learning models. The learning models could be a pre-compiled version so that they could be either transferred or ported to different devices or networks.

Google's main contribution to promoting edge-AI frameworks is by developing TPU-based hardware units. A TPU is a tensor processing unit that is custom-developed by Google to accelerate the performance of learning algorithms or inferences. The major difference between TPU and CPU is the approach to computing instructions, i.e., CPU computes instructions one after the other, whereas TPU can apply matrix-related operations in one stretch. Since TPUs are able to compute only matrix-related operations, they are not well-suited for general-purpose edge-AI computations as compared to CPUs or GPUs.

Owing to the restricted support for edge-AI applications, Google's Coral-based edge platforms are limited to specific applications such as object detection, pose estimation, image segmentation, and phrase detection. In [17], the authors have utilized coral edge TPUs to demonstrate a high-speed object tracking system. In edge TPUs, the specified applications or domain problems are input with matrices while performing large-scale parallel computations in edge-AI.

Deploying edge-AI workloads in Google-supported edge systems focuses on evaluating the training stages and testing inferences. The platform has to verify the connectivity between training nodes and formulate an appropriate networking topology for identifying the services quickly. The users of edge-TPUs at Google are provisioned with a large array of machine learning environments. For instance, TPUs, GPUs, and the combination of multiple varieties of devices are utilized in Google's edge-AI systems.

The execution of edge-AI applications using Google Edge platforms such as Coral follows a uniform pattern of steps such as collecting data, pre-processing data, training and developing models, deploying models to different hardware components, evaluating the performance of algorithms, assessing compatibility issues, etc. A few authors have developed a pipeline for edge-AI applications, considering AirFlow or MLFlow frameworks.

Recently, Google has introduced a new product named Gemini. This product supports edge-AI by providing multimodal assistance and a dynamic learning process with the local datasets of the users. There are several anticipated applications using the Gemini tool. Notably, applications such as automated mark evaluation, enlightened reasoning, and a 360-degree processing of multi-model inputs such as texts, audio, and video have been developed using Gemini. Also, in recent years, a number of extensions have been developed. For instance, the authors of [19] extended Gemini to include fault-tolerant mechanisms.

9.4.3 Amazon edge

Amazon is a widely known cloud service provider. Although the journey of the company began by selling books, its contribution to cloud technologies is remarkably appreciated among researchers and developers.

In short, Amazon has developed around 321 cloud-related products in 22 subcategories such as compute, storage, IoT, networks, etc. The widely utilized IaaS cloud service is the Elastic Compute Cloud (EC2) service. The cloud-based services of AWS are deployed in multiple geographical regions across the globe, i.e., AWS services are divided into seven geographical regions: Asia, Africa, North America, South America, Europe, the Middle East, and Australia. The purpose of dividing resources across regions is to enable redundant services and data availability throughout the lifetime of the services.

To support IoT-related services, AWS has developed an array of products, ranging from protocol assistance to software-related services. Notably, IoT-related services are broadly classified into three categories, as listed below:

1. Device level services: Products such as IoT core and FreeRTOS are designed to increase the working methodologies of connected devices. The core features of the IoT core are establishing connections between sensor nodes or gateways with cloud services using MQTT protocols. The publish and subscribe model is developed in this framework while sharing data between nodes. The system adopts the shadowing services of the remote sensor nodes in cloud environments.

   The other product that is specific to AWS is the FreeRTOS product. In fact, FreeRTOS is an embedded OS that is widely applied to microcontrollers such as ESPs. Being an open-source operating system, AWS has tuned the software version to the requirements of AWS-related devices. The OS includes memory management and other tasking concepts for executing parallel tasks on IoT-enabled devices.
2. Control-related services: The control-related services of AWS relate to managing IoT-enabled devices in an IoT-cloud application environment. These services include AWS Greengrass, Device Defender, 1-Click, etc. AWS Greengrass offers edge-supported cloud services. The cloudlets are hosted on remote edge-enabled devices, so the services can be developed offline with the local datasets. Hence, the privacy of user applications is protected from various cloud-related security attacks.
3. AI-related services: AWS has developed more than five services that attempt to forecast, predict, or classify data. Notably, the AWS IoT Analytics

product consists of functions such as data collection, data processing, data storage, data analytics using ML/DL algorithms, and notifications. The entire service bundle can be integrated into sensor-enabled environments to notify users with selective ML algorithms such as random forests, support vector machines, XGBoost, etc.

Ambarella CVFlow, an SoC device, delivers computer vision-related solutions with low power assistance. This device could be combined with AWS edge services to increase the privacy of local datasets. For instance, AI processing is accelerated in the SoC chip using neural network acceleration techniques that are integrated with an AWS Sagemaker tool [5]. The purpose of integrating Sagemaker into the SoC is to easily develop models, compress the models if they are required, and port them to the device. In fact, the tool assists developers by easily porting the compiled versions of learning models to the SoC.

9.5 Application-specific frameworks

Edge-AI frameworks have been directly applied, with more emphasis, to applications such as agriculture, industry, education, and healthcare. These frameworks in different application domains have specialized prebuilt libraries or solutions with distinct characteristics and requirements. This section explores the development of application-specific frameworks with more emphasis on domains and requirements.

9.5.1 Edge-AI in agriculture

The major focus of the edge-AI framework deployed in the agricultural domain is on improving farming situations or knowledge delivery mechanisms. A typical framework in agriculture will have the following components:

1. User-interface: An interface is a device that performs human–computer interactions. The user interface designed for agricultural farms has to be suitable for farmers and environmentalists. As farmers are often not well-trained to operate high-end user-interfaces or devices, the user-interface has to be user-friendly, and it has to enable all required features.
2. Sensor-enabled data automations: The framework will have provisions to collect sufficient sensor data using sensors that are distributed across multiple regions. The collected data has to be pre-processed in real-time or offline based on the needs of the application. Some pre-processing techniques, such as denoising, filtering, augmentation, etc, have to be carried out without much human intervention in such frameworks.
3. Data analytics: Depending on the processed data, models have to be developed, trained, validated, and tested for sensed data from agricultural farms In this module, several learning algorithms could be incorporated to deliver accurate predictions or classifications.
4. Notification: Finally, the framework has to deliver reports or insights from the future data. The framework has to support a wide variety of output formats for the notification of agricultural findings to users.

In fact, there are tens of thousands of applications oriented toward improvements in agricultural sectors. These applications could be developed under three broad categories: (a) smart agriculture, (b) sustainable agriculture, and (c) precision agriculture.

9.5.1.1 Smart agriculture

The primordial objectives of developing frameworks that enable smartness in agricultural edge-AI applications are: (i) to apply learning algorithms or intelligence to agricultural systems, (ii) to observe the impact of notification services levied by the associated frameworks, and (iii) to collect the history of previous decisions so that agricultural harvest times could be automatically determined by frameworks.

For instance, a 30 MHz company has developed a smart agriculture system using AWS IoT core and AWS Greengrass products from Amazon to remotely monitor the health conditions of indoor plantations.

In addition to developing novel learning algorithms, researchers have established smartness in agricultural farms by proactively calculating the anomalies while planting crops [11]. The crops or leaves of plants could be infected by insects or animals. This degradation in the future revenue of land is monitored and administered within a short span of time.

9.5.1.2 Precision agriculture

Precision agriculture or farming, aims at increasing the accuracy involved in the predictions or classifications of agricultural datasets. This means that a few robust learning algorithms need to be incorporated into the agricultural edge-AI frameworks to deliver sophisticated learning accuracy. Some notable metrics, such as accuracy, true positives, squared errors, etc, are attempted to improve the system.

To increase the accuracy involved in predictions and classifications, the framework that focuses on precision agriculture will enable the real-time collection of remote data from sensors. It has to deal with the dynamic nature of data evolving from heterogeneous systems. The authors of [16] have revealed an IoT-enabled edge-AI framework for precision agriculture.

9.5.1.3 Sustainable agriculture

If the precision of prediction accuracy is reasonably accepted in an agricultural edge-AI application, revenue will increase. It is also possible to detect any health-related concerns well in advance. Hence, sustainable agriculture focuses on three broad objectives:

- To increase the revenue involved in the production of agriculture using sophisticated ICT technologies, including edge-AI frameworks.
- To provide societal benefits by leveraging the utilization of artificial intelligence along with the assistance of edge-enabled solutions.
- To ensure healthy food production or farming is practiced for the sustainable growth of societies.

Sustainable agriculture is practiced at Google by incorporating the Google Earth engine platform, which collects GPS data on different landforms The major notion of Google's development of the Google Earth engine platform is to increase awareness of food scarcity and agricultural parameters that improve human revenue. Notably, a project developed using the Google Earth engine has enabled regenerative agricultural suggestions in Australia. For instance, the Regrow Company of Australia has showcased the availability of carbon sequestration of soils and gas emissions to environments [6]. Similarly, the probable yield that could be generated using rubber milk production could be predicted to increase the sustainability of agricultural farms.

The details of various applications targeting these categories are described in chapter 11.

9.6 Distributed federated learning frameworks

The edge-AI frameworks have been implemented in various ways while incorporating AI-related concepts into them. One of the methods is to develop a learning model using several decentralized approaches and regional datasets. Federated learning is an approach that aims at developing a single model with the assistance of several decentralized models. This section explores the federated learning-based frameworks and the inner details of the associated software components.

Broadly, the discussion in this section focuses on a few available federated learning frameworks that utilize edge-enabled data sources. Notably, the frameworks such as (i) federated learning, (ii) SplitFed, (iii) FedAdapt, (iv) FedlesScan, (v) Fate, (vi) OpenFL, (vii) tensorflow federated, (viii) FLAME, and (ix) Clara are discussed.

9.6.1 Federated learning

A traditional implementation of a federated learning framework consists of a master and multiple clients. The clients could be edge devices, mobile phones, laptops, sensor-enabled nodes, etc. The notion behind developing a master and a set of client nodes is to involve them in collaboratively developing a single global learning model: i.e., the master will share the weights of models and the clients will engage in processing the model based on the local data that are available from regional locations [14].

The training models from local sites are transferred to master nodes for consolidation or aggregation. During the process of consolidation, different techniques are adopted in the frameworks. For instance, averaging data models collected from multiple clients to create one single model is a consolidation approach in traditional federated learning. In fact, the clients and master nodes will have to continually work on the local weights and update the global weights to improve the accuracy of the learning models. In federated learning, gradients and updated weights are transferred to master nodes to compute the global weights in a periodic fashion until a specified accuracy is met.

In summary, the major steps involved in the naive federated learning framework are:

1. The master node is a server instance that submits the global model to identified client nodes.
2. The client nodes undergo learning based on the model and the local datasets. Accordingly, any updates on the weights of the learning parameters are observed.
3. Subsequently, the client nodes resubmit the updated weights to the master nodes.
4. At last, the master node consolidates the weights received from multiple clients to repeat the evaluations observed in the global learning model.

9.6.2 SplitFed

Combining split learning and federated learning emerges as a new framework named SplitFed [18]. The purpose of merging split learning and federated learning is to avail of the benefits of both methods.

It is a known fact that split learning is a distributed deep learning approach in which the computations relating to hidden layers are spread across the available computing nodes, including edge nodes; the cut layer is the layer that splits the computation across computing nodes; the layers are associated with agents so that the agents could communicate with one master node that performs the coordination of connecting the results from individual split computing layers. It is to be noticed that the master node will be responsible for coordinating the computations performed on client nodes.

Researchers have applied split learning in the context of back propagation and feed forward propagation in the past. The hidden layers of such algorithms are sequentially processed using client nodes. The gradients of the computations are further transferred to the master node through the agents of clients. Similar situations occur for back propagation or other learning algorithms.

The drawback of providing only the features of the split learning concept is the difficulty in handling the unbalanced data of clients. The clients could not attain the expected outcome due to the varying data components between them. Also, the second major drawback of split computing is due to the processing of all layers in a sequential manner. Owing to the sequential computations, the resources are not efficiently utilized when the entire edge-AI application is considered.

The combination of split learning and federated learning overrides the drawbacks of split learning processes, especially the sequential processing and unbalanced data situations. To do so, the split-fed frameworks partition deep learning algorithms at pre-defined cut layers. This step is similar to the split-learning process. However, instead of delivering the entire layer to clients, the SplitFed framework approach provides a portion of layers to clients and the other portion of layers to master nodes. The notion of engaging masters for layer computations is to avoid unbalancing the data of clients by averaging the models on clients and masters.

This means that the back propagation and feed forward propagation steps could be carried out in combination with clients and master nodes.

In the past, a few variants of SplitFed learning have been developed: i.e., aggregation of model weights is carried out in clients instead of master nodes in some implementations. This reduces the computational overhead that could be incurred by the master node otherwise.

9.6.3 FedAdapt

The FedAdapt framework is developed to adaptively decide on the cut layer of the split learning processes discussed in chapter 4. In split computing environments, the cut layers determine the stage at which to offload computations to clients [20]. However, there is no specific pattern to explain the split involved in such neural network environments. The clients receive a set of neural network layers without knowing the reason for delivering such computations. This approach could have several disadvantages:

1. The layers of the neural networks are split in a naive fashion. For instance, if there are 100 hidden layers, the framework could decide on splitting the layers into 10 equal subdivisions.
2. The layers are distributed based on the decisions of the reinforcement learning agents in the framework. Obviously, the agents could impact the performance of learning models. It is due to the fact that the weights of gradients in the neural networks have to be propagated between computing nodes that execute the learning algorithms.

The FedAdapt framework applies a reinforcement learning algorithm to offload tasks to edge devices. The idea behind developing a reinforcement learning algorithm for the offloading process of the framework is to prepare a sequence of learning decisions so that rewards can determine the offloading points or actions. Researchers have put forward ideas to improve the rewarding schemes or reinforcement learning steps of the FedAdapt framework.

There have been efforts to improve the offloading decision time by developing hardware accelerators. The purpose of the hardware is to accelerate the offloading strategy and improve the training time involved in the application. In a few works, researchers have applied quantization techniques to reduce the anticipated communication overhead between the clients and master nodes.

9.6.4 FedLesScan

Researchers at the Technical University, Munich, Germany, have been involved in developing a federated learning framework using serverless functions [9]. The FedLesScan framework benefits from the application of serverless computing concepts.

In general, a serverless computing environment is offered by public cloud providers as part of their novel infrastructure support without user provisioning or user management tasks. It is an execution model in which servers are not available unless there are triggers from clients. Based on the triggers, the functions are executed by identifying an appropriate server node from the pool of server farms.

According to the serverless computing concept, the servers are not assigned to execute the applications. This means that cloud-based applications have to search for cloud instances, such as VMs, dynamically. The servers are allotted by the infrastructure based on requirements and necessity. The server instances, at times, have to be powered on, booted, and service enabled for executing all these tasks. However, serverless computing has various advantages, such as (i) reduced overall costs of executing applications, (ii) negligible managerial tasks, (iii) high scalability features, and (iv) reduced resource utilization.

In general, the server instances could be bare metal devices or Docker containers. Sometimes, the server instances are not available, and they have to be powered, created, and booted automatically. This feature is referred to as cold-start instances of servers. Otherwise, the server instances are available, and they have to be identified from a pool of idle server instances. This feature is termed the warm-start instance of servers. Obviously, serverless computing provides a unique feature to minimize the power consumption of applications.

In FedLesScan, due to the inclusion of serverless computing in the framework, the computations between master and client nodes are triggered using serverless functions. The framework utilizes Kubernetes to provide a scalable environment between master and client nodes. It has an OpenWhisk cluster to impart function-oriented triggers. A MongoDB database is utilized in the framework to collect the details of client nodes and the availability of instances of them over a period of time. The global model is stored in a database so that clients can update the weights of the global model from time to time using the framework. Apart from these components, the framework has also included a security component for enabling authentication. In this way, only authorized clients are involved in updating the global model of the federated learning framework.

Although the entire computational requirement of a serverless framework and the energy consumption of the framework could be minimized, the complexity involved in the framework is too high. A few notable challenges that are identified in the framework are listed below:

1. Stragglers: The client nodes of a serverless federated learning framework named FedLesScan could lead to potential stragglers. There are several factors that determine the stragglers in the framework. For instance, heterogeneity of devices, data incompatibility, network failures, node incompatibility, etc.
2. Implementation complexity: The framework has to be deployed using two widely utilized software modules: i.e., Kubernetes and OpenWhisk. The OpenWhisk cluster has to be deployed on top of the Kubernetes cluster. This is one of the difficulties that users encounter with the framework.
3. Selection of strategies: The framework is not automated to determine the selection of clients or allocation strategies. For instance, the framework has a limited scope of flexibility in choosing different allocation strategies when compared to reinforcement learning approaches.

9.6.5 FATE framework

Federated AI Technology Enabler (FATE) is a federated learning framework that is oriented toward the industrial sector. The FATE framework is developed by WeBank's AI department. It focuses on developing a federated learning framework that supports edge-enabled devices, with more emphasis on security. Being the bank's innovation, the security features are by default included in the framework.

The framework offers users a federated learning environment to collaboratively develop learning models from the data residing on the local machines. The framework is implemented as a modular framework. This enables developers to scale the implementation based on the requirements and advancements. For instance, the security involved in edge-level computing units could be updated with novel security features or replaced with another security algorithm.

This tool is developed as an open-source solution based on the Python programming language. FATE is developed using WeBank Inc. It is designed using four layers: (a) computing and federation layer; (b) secure protocol layer; (c) federated ML layer; and (d) UI tools.

The computing and federation layer involves computing environments that deploy computing nodes across devices. The FATE tool utilizes the TensorFlow computing framework and the RabbitMQ approach [13] to enable the federation of learning environments among the distributed nodes. It supports various ranges of storage units, such as SQL, HIVE, HDFS, etc.

In order to enable security features, FATE embodies homomorphic encryption techniques. In these techniques, the original data is encrypted using specific algorithms before being transmitted across learning-computing nodes.

The Federated ML layer is the core layer of the FATE architecture. It involves a variety of ML components to support a horizontal or vertical federated learning process. The layer is responsible for optimizing the learning model that will be created by the regional datasets of the system.

The core components of the FATE tool are deployed using two popularly developed deployment frameworks, the Eggroll and Spark frameworks. This increases the utilization of the framework for various applications and distributed learning tasks in a federated manner.

FATE has multiple federated versions. One of the implementations is Federated ML. Federated ML is a library-based implementation framework that can be utilized to write edge-AI applications. The library has several algorithms, such as DataIO, Federated Sampling, Feature Scaling, and Hetero Secure Boosting. A list of encryption tools and statistic-based modules has been implemented in the framework for the convenience of edge-AI application developers. The framework enables a skeleton for developing newer algorithms or scale modules that can be applicable for edge-AI applications, in addition to the core application development based on federated learning. The core of the framework is to focus on security-related features that can enable secure communications between computing machines that perform the federated learning processes.

9.6.6 OpenFL framework

OpenFL, as the name indicates, is an open-source implementation of federated learning from the Linux Foundation. The implementation of OpenFL is based on a library style that is written in the Python programming language. The framework supports both machine learning and deep learning algorithms in a distributed fashion, such that computations are leveraged by multiple compute machines [10]. Additionally, the framework can be integrated with two existing ML/DL frameworks, such as Pytorch and TensorFlow, to create federated learning environments.

The core features of the OpenFL framework are: (i) workflow-based FL integration for applications, (ii) modular-based implementation, and (iii) easy adoption. The framework paves the way for edge-AI application developers to set up federated learning algorithms using the Director-Workflow feature. This feature can offer environments to setup long-lived algorithms for applications. Also, the Aggregator-Workflow feature of openFL paves the way for researchers to explore the verified learning models that are integrated in the FL environment.

9.6.7 Tensorflow federated

Tensorflow Federated is an open-source implementation of a federated learning framework that handles machine learning for distributed data. The framework provides a simulation environment for researchers to explore the performance or accuracy components of federated learning algorithms. Many researchers have observed the capabilities of novel learning algorithms without setting up real edge-enabled environments.

The simulation environment is built up in an API style with two levels of implementation: the core API and the high-level API. The core API permits users to study the inner features of a federated environment. In contrast, the high-level API is written to enable users to learn the impact of different federated learning algorithms without modifying the simulation environment that is setup for the learning processes.

Figure 9.5 lists the edge-AI frameworks that have been widely applied by researchers and discussed in the previous subsections.

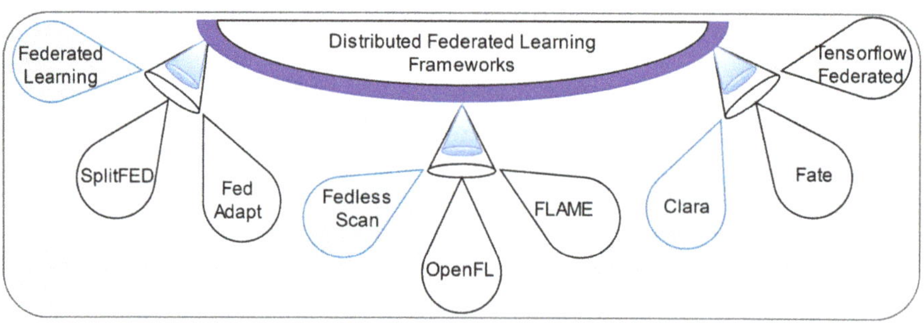

Figure 9.5. Most commonly utilized edge-AI frameworks.

9.6.8 FLAME tool

The FLAME tool was developed by Cisco Research. It aims to create federated learning models considering the privacy aspects of distributed locations [7]. The training dataset is localized while creating models. To do so, FLAME ensures that the global model is created based on the weights received from the learning models of the localized training dataset.

FLAME is implemented as a service on top of Kubernetes clusters. The services include ML-based learning tasks and management-related tasks. The ML-based learning tasks are framed to collect learning weights from distributed learning environments; the management-related tasks aim at improving deployments and scalability-related constraints by providing appropriate scheduling mechanisms. The tool includes an SDK to develop newer learning models and to design topologies considering the available computational nodes. The purpose of incorporating Kubernetes clusters is to enable easy deployment in any scalable environment. Currently, the FLAME SDK is written for Python programmers who have knowledge of the Keras and Tensorflow frameworks.

9.7 Conclusion

The existence of a number of pre-built libraries or solutions as edge-AI frameworks can encourage application developers to quickly start them. This chapter lists some frameworks that could be adopted in various application domains. The frameworks that focus on allocating resources, managing cloud services, utilizing application code snippets, and attributing federated learning aspects are discussed in the chapter. Readers, therefore, can look into the adoption of a framework among the list of available ones for their edge-AI application.

References

[1] Edge TPU inferencing overview : Coral–coral.ai 2023 (https://coral.ai/docs/edgetpu/inference/) (accessed 20 February 2024)

[2] Abirami S and Chitra P 2020 Energy-efficient edge based real-time healthcare support system *The Digital Twin Paradigm for Smarter Systems and Environments: The Industry Use Cases* **vol 117** ed P Raj and P Evangeline (Amsterdam: Elsevier) ch14, pp 339–68 *Advances in Computers*

[3] Alkohli 2023 Microsoft Azure Stack Edge Pro 2 overview: learn.microsoft.com (https://learn.microsoft.com/en-us/azure/databox-online/azure-stack-edge-pro-2-overview#use-cases) (accessed 20 February 2024)

[4] Alqerm I and Pan J 2021 Deepedge: a new QoE-based resource allocation framework using deep reinforcement learning for future heterogeneous edge-IoT applications *IEEE Trans. Netw. Serv. Manag.* **18** 3942–54

[5] Ayar E and Balasubramaniam D 2022 ML inferencing at the edge with Amazon SageMaker Edge and Ambarella CV25 : Amazon Web Services: aws.amazon.com (https://aws.amazon.com/blogs/machine-learning/ml-inferencing-at-the-edge-with-amazon-sagemaker-edge-and-ambarella-cv25/) (accessed 31 January 2024)

[6] Bajwa K and Kaplan L 2023 Google Earth Engine and sustainable agriculture : Google Cloud Blog: cloud.google.com (https://cloud.google.com/blog/topics/sustainability/cop28-google-earth-engine-and-sustainable-agriculture) (accessed 31 January 2024)
[7] Cho H, Mathur A and Kawsar F 2022 FLAME: federated learning across multi-device environments *Proc. ACM Interact. Mob. Wearable Ubiquitous Technol.* **6** 1–29
[8] Stack Edge 2024 Deploy AI and machine learning at the edge: Azure Architecture Center: learn.microsoft.com (https://learn.microsoft.com/en-us/azure/architecture/ai-ml/idea/deploy-ai-ml-azure-stack-edge) (accessed 31 January 2024)
[9] Elzohairy M, Chadha M, Jindal A, Grafberger A, Gu J, Gerndt M and Abboud O 2022 Abboud 2022 FedLesScan: mitigating stragglers in serverless federated learning *2022 IEEE International Conference on Big Data (Big Data)* pp 1230–7
[10] Foley P *et al* 2022 OpenFL: the open federated learning library *Phys. Med. Biol.* **67** 214001
[11] Gupta B, Madan G and Quadir A 2022 A smart agriculture framework for IoT based plant decay detection using smart croft algorithm *Mater. Today Proc.* **62** 4758–63 (International Conference on Innovative Technology for Sustainable Development)
[12] Kong X, Wang K, Wang S, Wang X, Jiang X, Guo Y, Shen G, Chen X and Ni Q 2021 Real-time mask identification for COVID-19: an edge-computing-based deep learning framework *IEEE Internet Things J.* **8** 15929–38
[13] Liu Y, Fan T, Chen T, Xu Q and Yang Q 2021 FATE: an industrial grade platform for collaborative learning with data protection *J. Mach. Learn. Res.* **22** 10320–5
[14] Ludwig H *et al* 2020 IBM federated learning: an enterprise framework white paper v0.1
[15] Ng W C, Lim W Y B, Ng J S, Xiong Z, Niyato D and Miao C 2022 Unified resource allocation framework for the edge intelligence-enabled metaverse *ICC 2022: IEEE International Conference on Communications* pp 5214–9
[16] Rahmouni M, Hanifi M, Savaglio C, Fortino G and Ghogho M 2022 An AIoT framework for precision agriculture *2022 IEEE Intl Conf on Dependable, Autonomic and Secure Computing, Intl Conf on Pervasive Intelligence and Computing, Intl Conf on Cloud and Big Data Computing, Intl Conf on Cyber Science and Technology Congress (DASC/PiCom/CBDCom/CyberSciTech)* pp 1–6
[17] Sengupta J, Kubendran R, Neftci E and Andreou A 2020 High-speed, real-time, spike-based object tracking and path prediction on google edge TPU *2020 2nd IEEE International Conference on Artificial Intelligence Circuits and Systems (AICAS)* pp 134–5
[18] Thapa C, Arachchige P C M, Camtepe S and Sun L 2022 SplitFed: when federated learning meets split learning *Proc. of the AAAI Conf. on Artificial Intelligence* **36** 8485–93
[19] Wang Z, Jia Z, Zheng S, Zhang Z, Fu X, Ng T S E and Wang Y 2023 *GEMINI: fast failure recovery in distributed training with in-memory checkpoints, SOSP '23* (New York, NY: Association for Computing Machinery) pp 364–81
[20] Wu D, Ullah R, Harvey P, Kilpatrick P, Spence I and Varghese B 2022 FedAdapt: adaptive offloading for IoT devices in federated learning *IEEE Internet Things J.* **9** 20889–901
[21] Xu S, Zhang Z, Kadoch M and Cheriet M 2020 A collaborative cloud-edge computing framework in distributed neural network *EURASIP J. Wirel. Commun. Netw.* **2020** 211
[22] Yang Y-T and Wei H-Y 2023 Edge–IoT computing and networking resource allocation for decomposable deep learning inference *IEEE Internet Things J.* **10** 5178–93

IOP Publishing

Edge Intelligence
Deep learning-enabled edge computing
Shajulin Benedict

Chapter 10

Orchestration platforms: computing continuum

Orchestrating resources and services in an automated fashion is one of the mandates of deploying edge-AI applications. Accordingly, researchers and developers have delivered methods and solutions in edge-AI frameworks or architectures. This chapter explains the works that focus on orchestrating resources or tasks for edge-AI applications. It explores the orchestration patterns and pros/cons of applications.

10.1 Orchestration and integration

The orchestration of resources and tasks to perform edge-AI tasks effectively is crucial for modern applications. Application developers are not proficient in knowing the heterogeneity of devices; they are not capable of handling the integration of devices that are involved in edge-AI applications. The orchestration or integration platforms are responsible for mapping appropriate resources for edge-AI applications.

10.1.1 Characteristics

The orchestration platforms should possess a few important characteristics for assisting application developers with edge-AI applications. The list of characteristics that a platform should provide is given below:

1. The orchestration platform should deliver autonomous decisions considering the inherent parameters of environments. The platforms include software and hardware components to perform automated decisions.
2. The platform has to learn the environment by monitoring the execution of applications in real time.
3. The orchestration platform has to provide appropriate workloads for the right volume of identified resources. If the platform is not well developed, the resources and tasks will not be mapped efficiently. This leads to poor performance.

doi:10.1088/978-0-7503-5593-3ch10

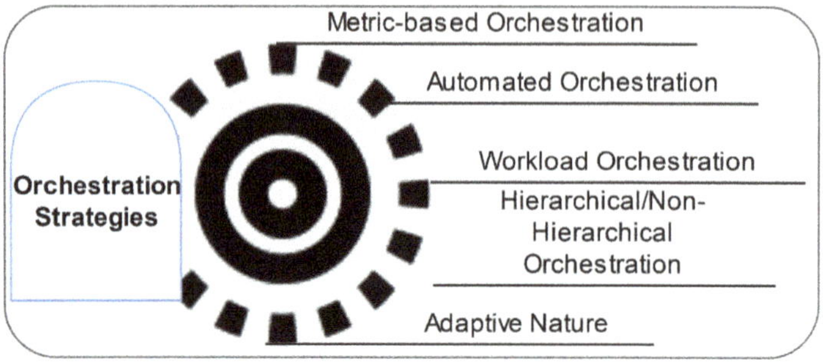

Figure 10.1. Holistic view about orchestration approaches in edge-AI.

In fact, it is very challenging to identify a holistic design or framework that incorporates all the required characteristics of orchestration platforms.

10.1.2 Approaches

Orchestration platforms for edge-AI applications can be developed with an emphasis on various factors, including workloads. For instance, the platforms could focus on orchestrating workloads instead of bringing together resources; the platforms could increase the parallelism or distribution of tasks by developing algorithm-specific modifications; the platforms could emphasize the delivery of resources, i.e., the hierarchical or non-hierarchical distribution of resources while solving edge-AI applications; some platforms engage in improving the adaptiveness of workload or resource orchestration; some platforms focus on providing resources or executing tasks in an automatic fashion; some platforms focus on sharing resources based on performance metrics, etc.

In subsequent sections, we will learn about the orchestration platforms that emphasize different perspectives. Figure 10.1 provides a holistic view of the orchestration approaches that will be discussed in the forthcoming sections.

10.2 Algorithmic/application orchestration

Algorithmic orchestration is an approach to coordinating and managing applications. In edge-AI domains, applications have certain objectives that can be split into smaller functions or modules. The orchestration of applications aims at combining multiple applications or synchronizing the capabilities of smaller tasks/functions to fulfill the required functional logics.

In most cases, the applications are represented in the form of services. The services deliver service-oriented implementations that serve certain purposes to fulfill the requirements of users. The idea behind orchestrating multiple applications or services is to provide multiple loosely coupled service instances without offering a direct integration of the objectives. In doing so, applications, especially in the edge-AI domain, could attain more flexibility when compared with the traditional approach of providing a direct integration platform.

The orchestration of algorithms or applications is carried out in a distributed service establishment pattern. The major reason for providing a distributed implementation is to appease the requirements of geographically dispersed, tiny edge devices.

For instance, applications employ microservices that are hosted on Kubernetes-based service orchestration platforms The notion behind utilizing Kubernetes or Docker Swarm platforms is to increase the scalability and manageability of service instances. [15]. Kubernetes or Docker Swarm platforms have several features, such as monitoring containers that host microservices, replicating resources based on user requirements, and migrating containers or services across multiple connected nodes of the platform.

10.3 Workload orchestration

Workload orchestration is the process of orchestrating tasks based on the specified flow of computing operations, considering the underlying computing resources. In contrast to application orchestration, workload orchestration considers resources and processes. In general, workload corresponds to the quantity of compute resources that are required to execute applications or services. TinyML-Ops [2] delivers an ML-based workload orchestration framework for data scientists.

In fact, the need for workload orchestration in the edge-AI domain became visible when industries associated with IoT attempted to automate services and manage configurations for tiny devices. It can be seen that the tiny devices have to be booted and enabled to execute applications. In some cases, the configuration settings need to be accomplished for the smooth transition of services.

A few edge-AI applications that are represented in the form of Directed A-cyclic Graphs (DAG) consist of a series of tasks that could be executed on parallel machines: i.e., the output of a set of tasks becomes an input for the other set of parallel or serial tasks. It is mandatory to coordinate these tasks, considering the underlying resources such as containers, VMs, storage units, etc.

Workload orchestration has to deal with security features associated with the hardware and software solutions. For instance, compute resources such as memory, CPU, GPU, and TPU have to be instantiated and employed for applications considering authentication features and avoiding security-related attacks.

A few researchers have developed a fuzzy approach to orchestrating workloads in edge-enabled systems [11]. In general, fuzzy logic corresponds to opting for multiple decisions instead of providing a complete truth or a complete false. The degree of decisions prompted by fuzzy logic offers multiple values. The concept of fuzziness was developed in the 1960s. It evolved from the fuzzy set theory of mathematicians.

The purpose of including fuzzy logic in edge-AI platforms is to handle multiple environments that are highly dynamic in nature. The application of fuzziness in decisions about handling workloads can harden the appropriate variables that increase businesses. Additionally, the precision of judgments could be improved amidst the potential uncertainty expected in environments.

10.4 Hierarchical versus non-hierarchical orchestration

Orchestrating resources or services on edge-AI platforms could be accomplished by establishing two functional taxonomies: (i) hierarchical and (ii) non-hierarchical. The hierarchical provisioning of resources for services or applications means layering decisions. If hierarchy is implemented in edge-AI platforms, the nature of decision-making varies as there are no single decisions. However, the management of decisions could be controlled based on the higher levels of decisions.

The hierarchical orchestration provides a clear distinction in delivering resources. The hierarchy could be enacted based on the operational domains, performance readiness, solvable conditions, etc.

The major advantages of providing a hierarchical orchestration approach are listed below:

1. Modular decisions: Decisions on allocating resources and coordinating services are grouped in a way that users can experience autonomous judgements within their jurisdiction. Also, on-demand policies could be incorporated within each hierarchical implementations.
2. Abstracting workloads: It is possible to abstract resources or their working conditions by developing abstractions. Similarly, the abstractions of services and applications could be created. The purpose of abstracting workloads is to quickly develop decisions and monitor the service-level conditions of edge-AI applications.
3. Debugging easiness: The hierarchical distribution of workloads or service instances can minimize the bottlenecks involved in identifying the bugs. It is often the case in software developments that the bugs could not be traced in a reasonable time, especially when distributed computing systems such as edge-AI frameworks are considered. With the application of hierarchies, we distribute the decisions in granular levels. This enables us to quickly debug the failures or potential bottlenecks in platforms.

10.5 Adaptiveness in orchestration

An important characteristics while orchestrating resources in edge-AI platforms is to adapt suitably based on the real-time parameters. The adaptive workload orchestration is a method incorporated in a few edge-AI platforms to modify the allotment of resources based on the current requirements of applications.

The continuous delivery of resources without considering the adaptive nature of workload requirements can be handled in a static or dynamic manner. The static delivery of resources is allotted by platforms or frameworks prior to starting the application. Once the application is started, the allotment of resources is not modified. The advantages of static allotment of resources are: (i) non-complexity; and (ii) timely completion of applications and services. That is, the resources are previously defined and allocated for running the applications. Thus, the orchestrated resources do not fail due to the non-availability of resources. Obviously, the completion time of such applications could be previously determined with greater accuracy.

However, the continuous delivery of resources in a static manner can impose several challenges:

1. The resources become underutilized due to overprovisioning. For instance, it might be required to start only two VM instances or Docker containers. However, as resources are orchestrated in a static manner, the overprovisioning method creates more than two instances that are required in a particular period of time. Obviously, it increases the cost involved due to the allotment of resources.
2. The resources could be underprovisioned in a static manner if they are not adaptively orchestrated. If the resources are underprovisioned, the services will not be completed within the prescribed time frame. The quality of service for specific applications can become much lower than the users feel otherwise due to the associated platforms and frameworks.

If an orchestration platform includes adaptiveness in the frameworks, the orchestration and monitoring of resources, considering the dynamic nature of input services, could be improved. To provide adaptiveness, the framework has to constantly collect a few performance-related parameters and hardware-specific information. Thus, the costs involved in executing applications or services for such edge-AI applications can be minimized as the framework involves cloud and edge-based devices. The network communications and the computational resources that are hosted in clouds could be minimized or adaptively improved in the adaptiveness-enabled orchestration systems.

The complexity involved in delivering resources might drastically increase when the orchestration happens in real time. There is also a direct proportion to the number of resources and services involved in each hierarchical layer in edge-AI systems.

10.6 Automation in orchestration

Automation in orchestration platforms is one of the prime qualities of the evolving tools in edge-AI domains. Automation minimizes the interaction of humans while executing applications or services. The main purpose of involving automation in services is to avoid redundant activities or methods throughout the implementation of applications.

For instance, on an agricultural farm, an automated orchestration would periodically identify the temperature, humidity, soil moisture, etc, from specified lands and deliver the data for pre-processing and actions without any human intervention. As all the associated tasks could be repeatedly involved in framing intelligence or the delivery of actions, these tasks could be automated for ease and faster computation.

The automation can reduce the delivery or update time involved in the hardware and software solutions. The software services will need to be updated at various levels. Notably, IoT-enabled devices have firmware and operating systems that could be updated based on requirements, such as security requirements. Thus, the update could be activated automatically to strengthen the security or performance features of applications and services. Similarly, the hardware portion of edge-enabled devices could be automated while engaging multiple hierarchical compute nodes.

Dynamic workflows are particular use cases in implementing automatic orchestration tools for edge-enabled systems. In dynamic workflows, tasks are created in a workflow fashion, i.e., they follow a DAG structure where the output of certain tasks becomes the input of a specified task. The workflows automatically accomplish a sequence of tasks without any human interventions on edge-AI platforms Increasing dynamism in such workflow execution environments is crucial in many modern edge-AI applications.

In recent years, researchers have developed several ML/DL pipelines to function as workflows [7, 9]. These workflows could be automated to increase the speed of code delivery in real-world working environments. For instance, AirFlow and MLFlow frameworks have been developed to create AI-based tasks using distributed edge-enabled applications.

Apache's AirFlow is considered to be a powerful data orchestration tool that can be incorporated into ML/DL-based systems [1]. The platform has given birth to Airbnb and, subsequently, has emerged as a powerful tool for Apache. The major outbreak of the AirFlow system is the provision for users to design their own data pipeline or orchestrating system. For instance, a DL engineer could develop multiple distributed neural network layers in the form of data pipelines. Due to its user-centric design, AirFlow distinguishes itself in the wide application of the system. The users could frame dependent tasks and formulate communication patterns in a programmatic fashion.

AirFlow or a similar workflow orchestration platform is applied in ML/DL to perform tasks as listed below:

1. Data collection: Data collection is the first step in the machine learning process. AirFlow or a similar orchestration tool could be utilized to collect data from various sources, including IoT sensors, in various data formats. For instance, the likes and dislikes of Twitter or YouTube users could be collected using specific Python-based APIs. The APIs could be designed using AirFlow's pipeline architecture.
2. Data pre-processing: In the pre-processing step, all irrelevant data can be removed or filtered. The redundant data can be removed and cleansed with the required amount of data, considering the computational capabilities.
3. Training—split computing: ML DAGs are created to train a model using datasets that are available in centralized locations or in distributed locations. This means that a few portions of neural network training processes will be carried out in different locations before collaboratively framing a trained model. DAGs could be designed in the form of pipelines to undergo training processes.
4. Validation: The purpose of validation in machine learning is to ensure that the model created during the training stage is suitable for performing future inferences. If the validation accuracy is not as expected, the model could be further trained and recreated. DAGs perform automated delivery of trained

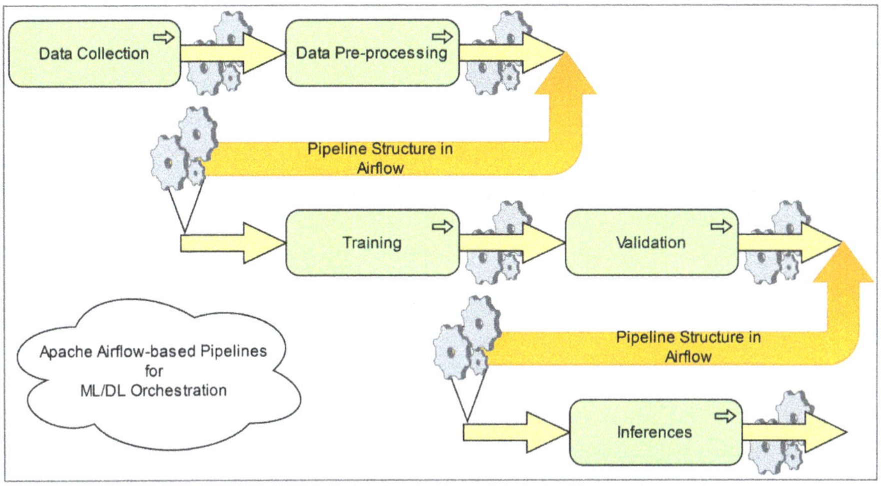

Figure 10.2. AirFlow-based pipeline architecture for edge/AI systems.

models in iterations so that human intervention is absolutely minimized. Obviously, the managerial efforts involved in the training, testing, and validation stages of any problem are drastically reduced using the pre-designed DAGs of AirFlows or similar tools.

5. Inferences: As a last step, DAGs can perform inferences and notifications by designing pipelines in the orchestration tools. In some cases, inferences could be reported in multi-modal formats, where each format could be designed as individual tasks in DAGs.

Figure 10.2 illustrates the flow of pipelines in the AirFlow-based edge-AI orchestration platform.

The major advantages of Apache's AirFlow are: (i) open-source implementation, (ii) highly scalable user-centric design, (iii) extensible, and (iv) automated repetitions. However, there are a few potential drawbacks to providing automated solutions using platforms such as AirFlow. Notably, the design could incur huge investments as skilled professionals have to elegantly plan the pipeline considering the industrial requirements and machines. Also, the interdependence of DAG tasks in AirFlow requires a huge correction mechanism while modifying any one task.

10.7 Metric-oriented orchestration

Performance metrics such as resource utilization, security features, fault tolerance, latency, network bandwidth, privacy levels, accuracy, and hardware cache could be efficiently utilized to automatically orchestrate workloads of edge-AI applications.

10.7.1.1 Resource-aware orchestration

In edge-AI systems, the applications have to collaboratively utilize resources from edge-to-cloud environments. The resources are either computationally capable, scalable cloud servers, or edge-based battery-operated devices. The orchestration platforms have to optimize the utilization of resources and channelize the tasks to appropriate resources in an automated fashion. The orchestration frameworks will be capable of distributing tasks across compute nodes, considering various parameters, including resource utilization aspects.

One approach to effectively utilize resources is to include resource scheduling policies or load balancing policies. The resource scheduling policies can focus on improving network utilization or compute node utilization strategies. The policies are often embedded into the orchestration frameworks to deliver tasks to appropriate resources without the intervention of humans.

10.7.1.2 Security-aware orchestration

Considering potential security attacks in the future, the orchestration framework could automatically consider security features such as access provisions to resources, controlled delivery of bandwidth, etc. In Docker instances, access to compute resources is provisioned only based on specific policies fixed in the agreements. The service-level agreements between users and resource providers are instantiated, considering security awareness.

In general, there are several security-related attacks which could be made on edge-AI systems. A detailed study of the potential security attacks is discussed in chapter 8. An orchestration platform should consider these security features and potential attacks before orchestrating tasks within the purview of compute resources.

10.7.1.3 Fault-tolerant orchestration

In many applications, the devices or frameworks are prone to failure. Edge-AI applications involve multiple edge-level compute nodes that are attached to sensors or actuators. Depending on the availability of sensor and actuator nodes, there are higher chances of powering OFF certain nodes. Obviously, the computational tasks stop executing instructions in the middle of the computations.

An orchestration platform in the edge-AI domain, therefore, has to be fault-tolerant. The advantages of designing an orchestration platform with fault-tolerant features are listed as follows:

1. Resilient to software and hardware failures: A fault-tolerant orchestration platform can withstand potential software and hardware failures. Applications, when executed on distributed edge-enabled compute nodes, have to depend on orchestrators to automatically resume tasks once they fail. The software could fail for several reasons, such as network communication failures, changing requirements, code optimization failures, agent failures, etc. Similarly, hardware issues happen due to disk failures, sensor-connected edge failures, circuit failures, etc.
2. Continued service: The services hosted on computing nodes continue to operate even if a service fails. The services that are hosted on VMs or Docker instances could fail due to planned reboots or unpredicted boot failures.

Also, the services could stop functioning due to algorithmic issues that are inherent to the system. The orchestration platform should consider these aspects of faults and plan scheduling the tasks to different compute nodes.

10.7.1.4 Latency-aware orchestration

Orchestration frameworks that are conscious of the latency involved in edge-AI applications are termed latency-aware orchestration frameworks. Latency is a metric that determines the delay in time to perform computations in edge-AI applications. The latency could be predominantly vivid in distributed computing applications such as computing compendium platforms which include edge, fog, and cloud computing clusters.

In network-based systems, latency occurs owing to packet loss or jitter. Packet loss is a phenomenon in which packets are not delivered to the intended MQTT brokers or subscribers due to network-related failures or similar issues. The jitter is a delay in packet transmission that can be visible in video-processing-based applications. The constant jitter between transmitter and receiver does not have a serious impact, especially in video-based applications, because of a constant delay in transferring packets. However, it is not acceptable for many other edge-AI applications. Jitters can have seriously impacts if they are not constant or predictable. The buffer size in receivers will have to be increased, apart from network bandwidth capabilities, to avoid jitter-related issues in edge-AI applications.

An orchestration platform has to consider the different jitter-related issues and re-route packets through the most suitable communication channels to avoid such latencies. If sensor devices have to submit data to cloud-based systems, the jitter is unavoidable in most cases. Also, the transmission medium between edge and cloud platforms varies in edge-AI applications. Therefore, an orchestrator has to consider all these pitfalls and opt for appropriate network channels while executing edge-AI applications.

In traditional networks, traffic shaping and bandwidth allocation strategies are followed to avoid network latencies. Traffic shaping is the process of limiting the transfer of data between compute nodes. It is a congestion management policy that is often practiced in networks to avoid delay jitters. The bandwidth allocation strategies are procedures to allocate network bandwidth in advance between communicating devices so that fair communication is delivered at a preferred time. The latency-aware orchestration platform could consider these traditional fault-tolerant features while orchestrating edge-AI tasks on highly dynamic distributed compute nodes.

The entire list of metrics that could be applied for tuning orchestration frameworks of edge-AI systems is depicted in figure 10.3.

10.8 Orchestration frameworks

Platforms include software and hardware components to orchestrate edge-AI tasks. In contrast, frameworks are software-only approaches to handling orchestrations in

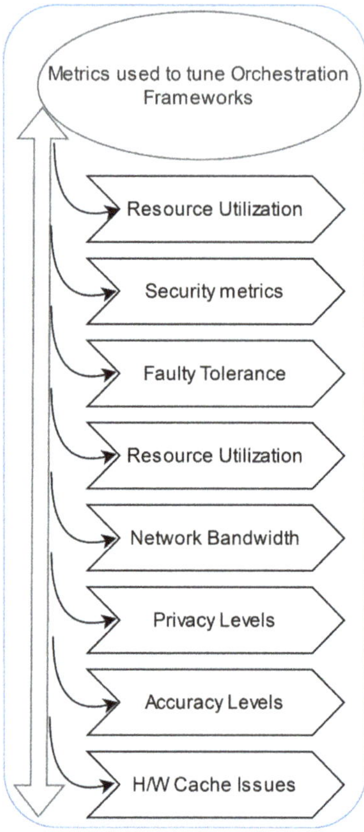

Figure 10.3. Metrics for tuning orchestration frameworks.

edge-AI solutions. This section explores the existing orchestration frameworks for developing edge-AI applications. The platforms that are discussed in this chapter are: (i) Oakestra framework, (ii) ORCH framework, (iii) Node-Red, (iv) KubeEdge framework, (v) Tiny-MLOps framework, and (vi) Pipeline frameworks. The frameworks provide a skeleton architecture to orchestrate services or functions and increase the utility of edge-AI applications.

10.8.1 Oakestra framework

The Oakestra framework is an edge orchestration framework that was developed at the Technical University of Munich, Germany. The framework offers a hierarchical orchestration of edge-AI applications [4]. The hierarchy is provisioned in the Oakestra framework to offer distributed decisions while orchestrating resources across regions.

The framework supports federated learning that could be applied in various application domains. The learning models based on deep learning are globally developed with local datasets and local models that are created in distributed locations. The models can be huge in size as BLOBs are supported in the framework while transferring them across distributed computing machines.

The framework can be integrated with specific service-level agreements. The users can prefer high-level SLAs to guarantee the distribution of resources for their tasks. The SLA option in the framework is provided to appease the requirements of users and providers, considering several other metrics such as bandwidth and latencies.

Another important feature of the framework is its lightweight implementation. The entire framework is supported by Docker-based containers with unified file systems. The notion of containers and scalability architectures is to port virtual instances across multiple devices. The lightweight implementation is well appreciated by several developers, users, and resource providers. In fact, the framework could be suited to heavy implementations using bare-metal server instances.

The major components involved in the framework are clustering components, namely the root orchestrator and the cluster orchestrator. The cluster orchestrator is a software entity that combines multiple clusters that are developed using Docker containers on edge or cloud-based compute regions. The cluster orchestrator component keeps track of available clusters and records node-related details on cloud-based databases. The framework has two categories of databases, namely serviceDB and nodeDB. The purpose of having a serviceDB is to record the available services and their status at a given time. Similarly, nodeDB is a cloud-based database that records the availability of nodes in a cluster.

The root orchestrator is a software component that collects all the details of available clusters that are utilized in the orchestration platform. The major purpose of the root orchestrator is to coordinate and manage the functionalities of the cluster orchestrators and their associated services or nodes.

The Oekestra framework has been applied to latency-aware applications. Most recently, researchers have applied the framework to an edge-AI application that involves cameras and sensor-capturing devices to perform augmented reality-based tasks. The researchers have demonstrated the application of the Oekestra framework while delivering authentic reality experiences to the users of the applications. In fact, the framework could be applied to various other edge-AI applications.

10.8.2 ORCH framework

The ORCH framework is a distributed orchestration framework for edge-enabled applications. The notion of developing such a framework is to handle the distribution of tasks on fog and edge devices. To this end, researchers have focused on monitoring the distribution of tasks on dispersed compute nodes, apart from the other crucial tasks such as selecting and deploying resources [13].

The ORCH framework is based on the FogFlow component. FogFlow consists of a user interface for designing flow-related IoT tasks. The framework includes master nodes and several client nodes. The responsibility of a master node is to coordinate all clients, or edge-level nodes. The master node decides on the offloading portions of computing tasks.

The master node of the ORCH framework includes a database to discover the availability of client nodes or worker nodes that are hosted in an edge-level computational area. The communication between master and edge nodes happens through brokers such as RabbitMQ. In principle, RabbitMQ follows the publish and subscribe pattern of communications. This means that the edge-level nodes submit data when it is available without waiting for the server to receive it. In contrast, the other subscriber modules, such as services or cloud nodes, receive the data when they need it through the communicating broker.

The activities are triggered using serverless functions in the framework. The purpose of serverless functions is to trigger instances based on computational requirements. The functions include the name of the function, the code base that needs to be activated, and event triggers that instantiate the requirements of computing nodes.

A few metrics that are used to decide the computational nodes in the ORCH framework are: (i) CPU utilization: the time when CPU is being utilized by executing applications, (ii) CPU error: the error that could not be determined by applications, (iii) CPU saturation: the time when CPU was overutilized and throttling happened due to the cause, (iv) memory utilization: the memory in GB or MB utilized for executing applications, (v) memory saturation: this determines the percentage of memory that is utilized in an application, and (vi) memory error: the amount of memory that caused failures while executing an application.

The major aim of the ORCH framework is to identify an appropriate edge computing node while executing serverless functions in a distributed environment. The whole process of the framework is to comprehensively identify the nodes while monitoring them periodically using monitoring tools, including Prometheus [12]. The master node of the framework centrally decides on the edge nodes, considering the availability of required packages or working environments for executing the applications.

10.8.3 Node-RED framework

The Node-RED framework is a graphical tool [8] developed by IBM in early 2016. It can be utilized to pictorially represent tasks and associated communications. The framework offers a browser-supported graphical interface to drag and drop different computing components and communication systems through its flow editor. The APIs are automatically inserted based on the framework. The framework has been successfully deployed in several IoT applications. However, it has been widely deployed in IIoT environments [6].

The computational requirement of the Node-RED framework is quite low when compared to several other computing nodes. For instance, the framework applies lightweight containers and node.js features for the quick deployment of services. As the framework is based on node.js, it can be extended based on the support provided by the online supporters and node.js developers. The framework could be deployed on local devices such as laptops, on edge-level devices such as Raspberry Pi boards, and on cloud servers.

Researchers have extended the features of Node-RED with a specific flow editor to orchestrate edge-based services. The Node-RED workflow manager of the research work has amalgamated several microservices to increase heterogeneity, dynamism, and discovery services for edge-enabled applications. The Node-RED-based workflow manager tool has two major components:

1. Process manager: The purpose of the process manager workflow manager is to start, update, and stop processes by creating end points for microservices that are oriented toward IIoT. In addition, the manager identifies tasks that need to be executed with appropriate resources. Once all tasks are defined, the process manager utilizes the task executor to execute tasks.
2. Task executor: The task executor gets the requests from the manager and executes them. At times, the task executor orchestrates the number of tasks that resemble each other. In doing so, synchronizing tasks and delivering tasks with increased QoS are possible in IIoT environments.

Node-RED has become popular in IoT cloud environments as it is offered by cloud giants such as Amazon and Microsoft. In Amazon AWS environments, users could provide a few configurational settings, such as the base node.js package version and Node-RED commands, to start services. It could be executed along with AWS Elastic Beanstalk to increase the availability of cloud-based services. The framework could be enhanced using AWS Greengrass and the cloudlets of Beanstalk to extend the services to edge-level nodes for any application domain, including IIoT.

10.8.4 KubeEdge framework

In general, Kubernetes is an orchestration platform that has been widely applied in OS-level virtualization environments such as Docker instances. The containers are quickly ported across computing nodes in online mode using Kubernetes. Being open source, the Kubernetes orchestration platform has been commercially deployed in several cloud-based applications.

The KubeEdge framework attempts to deliver a similar kind of orchestration service across edge-enabled computing nodes for containerized applications. The framework is based on the Kubernetes framework, with the main focus on edge-based, computationally challenging nodes. The major purposes of offering such a framework are to increase autonomy, extend computations to the edge, ensure seamless cloud-edge orchestration, and provide low-cost solutions. The framework is supported by appropriate APIs to provide container orchestrations [10].

The framework is hosted on a combination of edge and cloud infrastructure. It includes two major software components, termed CloudCore and EdgeCore. It is similar to the master and slave nodes of the Kubernetes framework. The CloudCore is hosted on cloud servers; whereas, the EdgeCore is hosted on one of the edge nodes. The purpose of CloudCore is to host master-level services on cloud servers so that the EdgeCore nodes can serve as worker nodes.

The CloudCore device includes edge controllers to restrict edge-level connections and provides access to Kubernetes. The connection of CloudCore with EdgeCore paves the way to extend connectivity to edge-level compute nodes. Most of the tiny devices, including sensor-based edge devices, are connected to the EdgeCore component using the publish/subscribe model. In such cases, the EdgeCore component serves as a Mosquitto agent so that the tiny devices can connect to the component using APIs.

All these components function using lightweight Docker-based implementations, as they are functionally based on the Kubernetes architecture. The containers are configured using specialized networks with unique IP addresses for each host-level worker node. They are represented as PODs in the KubeEdge framework. The framework is fine-tuned for orchestrating cloud and edge-level nodes so that edge-AI applications could be developed using the framework. The focus of the KubeEdge framework is to increase the reliability and scalability challenges of cloud-edge devices, with more emphasis on orchestrating computationally constrained devices.

10.8.5 Tiny-MLOps framework

MLOps is a machine learning operation that automatically deploys models to production environments. The concept of MLOps was developed from DevOps, where the time utilized for the software life cycle has decreased. DevOps includes development and operations in one place. The developers and operators had the opportunity to systematically interact and produce software. Hence, the applications were developed, versioned, and deployed with ease.

The concept of MLOps follows the principles of DevOps, with more emphasis on machine learning. Here, the machine learning models are created and deployed with a tight integration of developers and operators. The MLOps processes involve the following steps:

1. Data analysis: The input data and the format of the data are assessed in this stage of MLOps. The output of exploration leads to reproducible and editable data for further processing with machine learning algorithms.
2. Data processing: During this stage, the ML engineers attempt to learn the features involved in processed data. The notion of feature engineering for data is to select and transform the most important features of the data. This process can be supported in the later stages of inferencing, apart from the training phases.
3. Model development: The model development stage of MLOps is considered to be the most influential step for MLOps. In this stage, the mathematical models are created considering the extracted features of the data. The models differ based on different business goals that need to be formulated for different application domains such as agriculture, healthcare, transportation, etc.
4. Training phase: In this phase, the MLOps deliver models using specific tools and techniques in a test-bed-based production environment to fine-tune the model. The training phase needs the special attention of feature engineering

based on the environment in which the model needs to be deployed. The training phase needs to be iteratively handled within the automated pipeline of the MLOps framework.
5. Review and deploy: The developed ML is deployed to the real production environment after being tested and verified by the production team of engineers in an organization. It is the second part, the operations module, of the DevOps architecture.
6. Inference and monitor: The inferencing and monitoring stages resemble the relevance of the created model in the production environment. During this stage, MLOps can make suggestions to update the feature engineering stages or model creation stages in further iterations of the MLOps.

The entire MLOps stage is developed using specific tools and techniques such as MLFlow, CometML, MetaFlow, Kedro, Data Version Control, Cortex, etc.

In Tiny-MLOps, researchers have extended the pipeline stages of MLOps that consider the orchestration of ML tasks to resource-constrained tiny devices [2]. It is a known fact that large-sized models cannot be accommodated in tiny devices such as the Raspberry Pi. Hence, TinyML-Ops emerged for edge-enabled AI applications.

The Tiny-MLOps includes *Adapt and Optimize* stages between the model development and the training phase of the previously discussed MLOps. In thc *Adapt and Optimize* phase of the Tiny-MLOPs, the model is modified considering the availability of memory and CPU power in a battery-operated or resource-constrained device. In this regard, the model is tuned to fit within the limitations of the underlying devices. For instance, a Raspberry Pi device could execute MB-sized models when compared to a Jetson Nano with GB-sized models.

10.8.6 Pipeline framework

Pipelines are processes that follow a particular sequence of operations with constant flow, one after the other, without the intervention of ML engineers. MLOps deliver a sequence of operations in a particular flow where the operations could be iterated between developers and operators in an organization. In contrast, pipelines deliver end-to-end orchestrations.

The pipeline-based orchestration frameworks focus on three subtopics:
1. Data pipeline: In the data pipeline, the processes relating to data fetching, feature engineering, format modification, raw data processing, etc, are automated. In some cases, the pipeline has to deal with batched data whereas in other cases, the pipeline has to handle streaming data.
2. Model pipeline: In the model pipeline, model creation and steps relating to model tuning are automated. The model tuning includes hyper-parameter tuning and model delivery-related tuning processes.
3. Production pipeline: In the production pipeline, the processes that are responsible for delivering models to production environments are handled in a sequence of operations. Such pipelines include several fine-granular

software components for instantiating ML-related tasks on edge-AI platforms For instance, recording feedback from users or developers could be involved as a proactive task in the production pipeline.

Pipeline-based frameworks are most suited to several orchestrations' requirements for edge-AI applications. Notably, processes such as data collection, processing, training, validating, and testing in a production environment could be delivered in the form of pipeline architectures. It is expected that pipelines could improve the performance of model creation and deployments [3].

There are a few pipeline frameworks that are applied to orchestrating ML operations. Among them, a few frameworks extend their features to orchestrate edge-enabled devices and their associated edge-AI operations. In this subsection, we explore the widely utilized pipeline-based orchestration frameworks for AI applications.

10.8.6.1 KALE

KALE is an orchestration-based pipeline framework that automates ML processes [5]. The ML engineers are not very involved in various stages of model creation and deployment. This reduces the risk of engaging costly ML engineers during the ML workflow process. In this way, the framework allows users to develop the pipeline once with multiple instances of iterative operations using pipelines.

The core feature of KALE is to provide a reproducible environment using pipelines. The workflow is designed such that workflows are repeatedly utilized with varying inputs and outputs. Secondly, the KALE pipeline framework allows automation by offering end-to-end orchestration. Hence, the time to market is minimized, i.e., the production stage and the development stage are tightly coupled. Also, the maintenance cost is minimized due to the iterative production without the intervention of ML engineers. This requires the framework to proactively design the anticipated software components in the orchestration pipeline well in advance.

10.8.6.2 MetaFlow

MetaFlow is a pipeline-based orchestration framework that handles a large volume of data using sophisticated existing AI-based frameworks. MetaFlow is written as a Python library so that edge-AI developers can integrate their applications.

The pipeline tasks such as data processing, modeling, deploying, orchestration, and execution are elegantly handled using the MetaFlow library. It supports many other tasks, such as creating workflows, visualizing workflows, tracking their development, etc, during the development stage of ML-related applications. Additionally, scheduling executions on different distributed compute nodes and monitoring the execution of edge-AI tasks are automated in the pipeline framework of MetaFlow.

In fact, several organizations and ML enthusiasts have widely utilized the framework owing to its support for cloud giants such as Amazon and Google. Additionally, it could be easily deployed on any Kubernetes cluster. This increases the utility of the framework by extending the available clusters for ML-related operations.

The MetaFlow framework can be designed as DAG tasks in a graphical format. The visual representation of the framework can offer a linear transition between edge-AI tasks. An edge-AI application developer could utilize the framework with several tasks, such as optimizing the resources of edge nodes that are linearly self-contained within the two mandatory tasks of the framework: (i) `start`, and (ii) `end`.

10.8.6.3 AirFlow orchestration

The AirFlow orchestration tool is developed by the Apache Software Foundation. This orchestration tool is an open-source implementation that can automate ML tasks in a distributed manner. The AirFlow orchestration tool is designed in a modular fashion, such that many AI enthusiasts have utilized it to iteratively handle a large volume of data for ML-related tasks.

The AirFlow tool can be easily integrated with cloud-based services such as AWS services or GCE services to augment and process data by creating pipelines. The data pipelines are created as DAGs so that tasks are automatically scheduled. This orchestration tool follows two approaches to activating DAGs:

1. Static DAGs: In static DAGs, the tasks are statically defined for ML jobs. For instance, collecting data from specific industrial plants or agricultural farms could be designed as one task of a DAG. Similarly, all other tasks in a pipeline could be previously designed and activated to iteratively rerun without the intervention of ML engineers.
2. Triggered DAGs: In triggered DAGs, the tasks could be activated dynamically based on the triggers delivered from specific tasks of DAGs. Being a modular design, the modules of AirFlows could be integrated in a timely way using distributed computing nodes. Hence, such DAG designs of AirFlow orchestration platforms are beneficial to edge-enabled AI applications. Additionally, it is possible to decouple certain tasks from DAGs based on the triggers mentioned in the business logic of the previous tasks.

The pipelines of edge-AI tasks in AirFlow are written in the Python programming language. Hence, it is a user-friendly tool for many ML engineers. It has three major components:

- Visualization UI: This is the user interface. It is responsible for visualizing the DAGs of applications. The DAGs are represented using Python programming languages. They are compiled and visualized using the visualization tool in AirFlow.
- AirFlow workers: The workers are compute nodes that could be edge-based power-constraint devices. The duties of the workers are to report the status of the assigned ML jobs. In addition, the worker nodes have to compute the tasks delivered by the AirFlow scheduler from specific queues.

- AirFlow webserver: The AirFlow webserver contains scheduling-based services that investigate the database and take decisions, such as delivering tasks to a particular worker node or activating specific tasks in a DAG.

10.8.6.4 ZenML

ZenML is a data pipeline that can orchestrate edge-enabled devices to perform AI for applications [14]. ZenML is an integration-assistive pipeline framework developed by ZenML Inc. The idea of ZenML is to provide zero management effort to users while integrating resources or data pipelines for performing AI-related tasks.

ZenML has a few pre-defined infrastructural stacks based on GCE, AWS, and Azure platforms The stacks have information about configurations and settings that are uniquely defined for these cloud platforms.

Similar to the other orchestration pipelines, ZenML has a DAG visualizer that represents the ML tasks in a visual form. On clicking each node of DAG, the entire code snippet of the task will be visible to the developer. The developer can drag and drop the previously available tasks for ML application development. Also, the user could modify them or extend the code snippets to integrate the tasks with the other DAG tasks with ease.

The tool is designed to easily integrate the ML logic with cloud environments. For instance, the DAGs could be integrated with S3 buckets on AWS using the APIs of ZenML. The APIs are written in a Python code base, which makes them user-friendly for many data scientists or ML engineers.

The attraction of ZenML is its simplicity. Being an industrial endeavor, the ZenML framework is planned such that the pipelines could be extended for various working domains. To do so, ZenML is developed in the form of extensible modules. Another major attraction of the framework is the availability of several standard templates, considering the requirements of edge-AI enthusiasts. The templates could be easily ported to the development environments for customization to meet the needs of users.

The categories of orchestration frameworks that are widely represented in the edge-AI market are illustrated in figure 10.4.

10.9 Integration platforms

The orchestration platforms focus on automating logics involved in various DAG pipelines of edge-AI applications. They do not focus much on the connection between devices and automation systems: i.e., the orchestration platforms deal with the control of business logics involved in DAGs.

In contrast, the purpose of integration platforms is to ensure a smooth transition of data between connected devices and pipeline architectures. For instance, APIs, web services, message queues, or associated services could offer a common data delivery format that could be easily integrated between DAGs. It should be noted that the integration platforms do not modify any inner functionalities of DAGs.

It is expected that the integration platforms should support a wide variety of data formats and access options. Additionally, the integration platforms should diligently

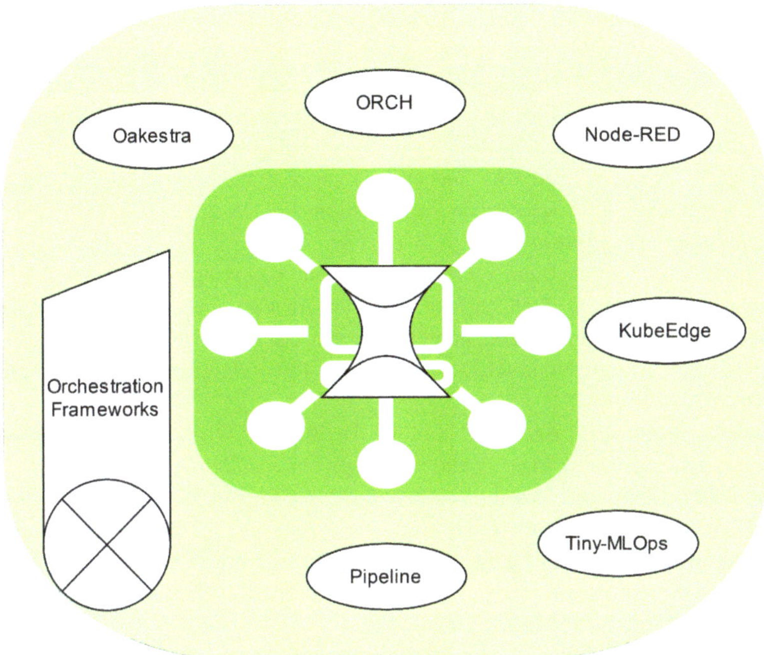

Figure 10.4. Most popular categories of edge-AI orchestration frameworks.

handle security-related issues and proactively avoid any security attacks. Obviously, the costs involved in such integration platforms could dramatically increase over time.

10.10 Conclusion

Orchestration is a key to automating services, including edge-AI services. This chapter delivers platforms that can orchestrate edge-AI applications. It projects the ways in which orchestration is adopted in the edge-AI domain. Notably, orchestrating workloads or algorithms, automating services, adapting ecosystems, etc, are studied. Additionally, the capabilities of the most notable orchestration frameworks to perform edge intelligence are explored.

References

[1] Airflow 2024 What is Airflow?—Airflow Documentation (https://airflow.apache.org/docs/apache-airflow/stable/index.html) (accessed 31 January 2024)

[2] Antonini M, Pincheira M, Vecchio M and Antonelli F 2022 Tiny-MLOps: a framework for orchestrating ML applications at the far edge of IoT systems *2022 IEEE International Conference on Evolving and Adaptive Intelligent Systems (EAIS)* pp 1–8

[3] Baresi L, Yi Xian Hu D, Quattrocchi G and Terracciano L 2023 NEPTUNE: a comprehensive framework for managing serverless functions at the edge *ACM Trans. Auton. Adapt. Syst.* **19** 1–32

[4] Bartolomeo G, Yosofie M, Bäurle S, Haluszczynski O, Mohan N and Ott J 2023 Oakestra: a lightweight hierarchical orchestration framework for edge computing *2023 USENIX Annual Technical Conference (USENIX ATC 23)* (Boston, MA: USENIX Association) pp 215–31

[5] Kale 2020 GitHub—kubeflow-kale/kale: Kubeflow's superfood for Data Scientists—github.com (https://github.com/kubeflow-kale/kale) (accessed 20 February 2024)

[6] Larrinaga F, Ochoa W, Perez A, Cuenca J, Legaristi J and Illarramendi M 2022 Node-red workflow manager for edge service orchestration *NOMS 2022-2022 IEEE/IFIP Network Operations and Management Symposium* pp 1–6

[7] Sundararaman S, Subramanian S, Ghanta S, Sridhar V, Arteaga D, Khermosh L, Roselli D, Talagala N and Luo Q 2018 usenix.org (https://www.usenix.org/system/files/conference/hotedge18/hotedge18-papers-talagala.pdf) (accessed 31 January 2024)

[8] NodeRed 2024 About: Node-RED—nodered.org (https://nodered.org/about/) (accessed 31 January 2024)

[9] Panchal D, Baran I, Musgrove D and Lu D 2023 MLOps: automatic, zero-touch and reusable machine learning training and serving pipelines *2023 IEEE International Conference on Internet of Things and Intelligence Systems (IoTaIS)* pp 175–81

[10] Phuc L H, Kundroo M, Park D-H, Kim S and Kim T 2022 Node-based horizontal pod autoscaler in kubeedge-based edge computing infrastructure *IEEE Access* **10** 134417–26

[11] Sonmez C, Ozgovde A and Ersoy C 2019 Fuzzy workload orchestration for edge computing *IEEE Trans. Netw. Serv. Manag.* **16** 769–82

[12] Tiivistelma 2019 Distributed orchestration framework for fog computing—LUTPub—lutpub.lut.fi (https://lutpub.lut.fi/handle/10024/160537) (accessed 31 January 2024)

[13] Tocze K and Nadjm-Tehrani S 2019 ORCH: distributed orchestration framework using mobile edge devices *2019 IEEE 3rd International Conference on Fog and Edge Computing (ICFEC)* pp 1–10

[14] ZenML 2024 Create an ML pipeline—docs.zenml.io (https://docs.zenml.io/user-guide/starter-guide/create-an-ml-pipeline) (accessed 20 February 2024)

[15] Zeydan E, Mangues-Bafalluy J and Turk Y 2021 Intelligent service orchestration in edge cloud networks *IEEE Netw.* **35** 126–32

Part IV

Applications

IOP Publishing

Edge Intelligence
Deep learning-enabled edge computing
Shajulin Benedict

Chapter 11

Edge-AI applications

Edge-enabled AI applications have been widely applied in various domains such as industry, healthcare, education, forensics, robotics, etc. This chapter explores the importance of edge-level AI for such application domains.

11.1 Applications

Applications are software components that are designed to assist users or other software modules. The applications serve certain user-defined objectives that mimic the requirements of real-world use cases. The application of edge-level intelligence in combination with cloud systems has increasingly evolved in recent years owing to technologies such as IoT and ML/DL. There are several domains in which edge intelligence could be applied: (a) healthcare, (b) manufacturing, (c) industries, (d) agriculture, (e) forensics, (f) logistics, and (g) robotics. The following sections will dive into the potential applications and probable use-cases that require edge-level intelligence for applications.

11.2 Edge-AI for healthcare

Healthcare applications are significantly increasing day by day because people wish to live longer. In fact, there are several innovations to protect ourselves from potential healthcare risks if appropriate data analytics is followed. For instance, a patient could proactively take medicines if the health-related data pinpoints near-future health hazards using AI technologies. This section explores the potential use cases of applying edge-level intelligence to the healthcare sector.

11.2.1 Remote healthcare monitoring and prediction

Remote healthcare monitoring and prediction have enormous applications in the COVID era. The outbreak of the COVID pandemic resulted in many chronic patients staying at home and not visiting doctors for many months, who were

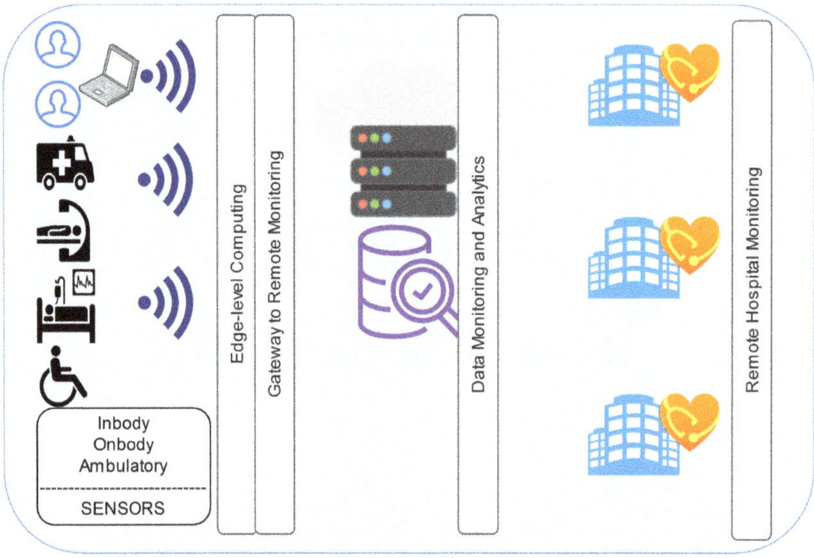

Figure 11.1. Remote healthcare monitoring: an example.

suffering from asthma, diabetics, cardiac problems, etc. This increased the application of remote healthcare monitoring using sophisticated technologies, including IoT.

11.2.1.1 Remote monitoring

Remote healthcare monitoring is a healthcare application that involves patients, doctors, hospitals, and organizations to provide remote healthcare consultation and monitoring [3]. The healthcare monitoring system includes devices such as mobile phones, sensor-enabled devices, personal gadgets, and some hospital equipment to provide sophisticated health-related checkups, monitoring, or predictions to remote patients.

For instance, mobile phones could be used to get appropriate remote healthcare consultations or diet advice from connected hospitals or doctors. In some cases, remote healthcare monitoring is useful to remind patients to correctly take their medicines. It is always possible that chronic patients either forget to take tablets or take their medicines multiple times, forgetting that they had already taken their medicines.

Figure 11.1 illustrates the architecture diagram for remote healthcare monitoring applications where edge-level computing of in-body sensors or on-body sensors is carried out. The architecture analyzes data with the assistance of remotely connected hospitals.

11.2.1.2 Prediction

Apart from remotely monitoring and notifying the health status of remote patients, it is also beneficial to forecast the future impact of the health conditions based on the monitoring data. For instance, if a patient is constantly registering blood pressure levels over 140, the remote healthcare system could forecast the possible health issues. It could also advise the patients on appropriate changes to their diet.

Prediction of health conditions could be obtained using prediction algorithms such as Support Vector Machines, Random Forests, Regressions, etc. A few deep learning algorithms could be applied to study the nightime sleep issues of patients. Notably, a diabetic patient could be injected with the right volume of glucose during the night by monitoring the sugar level and predicting sleep activity using Deep Learning algorithms.

Engaging edge-AI in medical data, including images, increases the chances of detecting health-related issues. For instance, CNN-based algorithms could be utilized to detect the types of skin-related cancers. The edge devices can internally process images and identify the type of skin cancer based on several parameters, such as skin color, infected region size, spreading nature of the skin, etc. Apart from being a specialized tool for early detection of health issues, it can become a prominent indicator for doctors while performing remote monitoring or diagnosis of patients.

11.2.1.3 An architecture

The remote healthcare monitoring system requires edge-level computations, as personal data should be protected by public parties. Although the majority of the services are based on the cloud, the data has to be prevented from being exposed to public repositories. Hence, the architecture that deals with the remote healthcare monitoring ideas has to be highly secure.

The remote healthcare monitoring system should include the following entities:

1. User-interface: A robust user interface that has 24/7 software bots to automatically share appropriate preliminary information and that suggests responses in a user-friendly manner is required in the healthcare monitoring system. The user interface supports virtual reality-based solutions that could be hosted not only on cloud systems but also on edge-enabled patient sites.
2. Sensors or end devices: Remote healthcare monitoring includes multiple types of sensors and edge-level compute nodes. Sensors such as temperature, ECG, heart rate sensor, etc, have to be periodically monitored depending on the nature of the applications. The sensor data needs to be pre-processed using edge-level computing nodes so that the data is not transferred to third-party compute nodes.

   Healthcare monitoring devices are categorized as in-body devices, on-body devices, mobile devices, ambulatory devices, and hospital devices. The purpose of in-body devices is to automatically capture or activate certain sensory data. For instance, sensor-enabled devices that are placed inside the body to continually report health conditions are termed in-body devices. The in-body sensors are often required by chronic patients, e.g., glucometers can be placed inside the body to automatically monitor the blood sugar levels of patients; the on-body devices are sensor-enabled devices that serve as wearable devices, consumables, fashion galleries, and assistive devices.

The on-body devices can help not only chronic patients but also users who like to check the status of their health at a specified time interval. Mobile devices are sensor-enabled medical devices that are not fixed to a geographical space. Ambulatory devices are IoT-enabled sensor/actuator devices that are fixed in ambulances to treat remote patients who need special attention prior to visiting hospitals. Hospital devices are devices that are available in hospitals. These devices are numbered and located based on geostationary information.

3. AI components: The data collected from various sensor-enabled devices in a remote healthcare monitoring system includes AI-based algorithms. These AI algorithms are utilized to either predict or classify data based on features identified in datasets.

11.2.1.4 Priority monitoring

Based on the health monitoring conditions and the availability of resources, the remote health monitoring system has to prioritize tasks. For instance, a periodic remote monitoring service should not require many resources for a normal user. Instead, it has to deliver resources to a critically ill patient. This reveals that the remote monitoring tasks need to be queued up for certain tasks.

Accordingly, the tasks involved in the remote monitoring system are classified as: (i) critical tasks, i.e. tasks that need to have an immediate attention, (ii) periodic tasks, the remote health monitoring tasks that have to be constantly activated over time, (iii) preventive tasks, tasks that are expected to detect cancer early or potential health risks using sophisticated AI algorithms, (iv) federated tasks, tasks that need to be collaboratively handled by multiple compute nodes or organizations, (v) alarming tasks: the healthcare-related tasks that have to provide warning messages or notifications, and (vi) educational tasks which are the tasks that provide some insights about health risks if some habits are not followed on time. Hence, priority-based monitoring of tasks in a remote healthcare monitoring system enable actuating tasks depending on the priority or importance of tasks and the availability of compute resources.

11.2.2 Personalized medicine

Personalized medicine can offer more accurate medication tailored to the need for medicines. The objective of having personalized medicine is to increase the accuracy of treatments and improve the well-being of patients. There are cases where doctors prescribe medicines that are generic for diseases. This happens due to the medical treatment provided by less experienced doctors. Specialist doctors are then overloaded due to having to deliver personalized medicine to patients due to a large number of patients in hospitals.

With the remote monitoring solutions, patients could get more personalized attention as they will be assisted by IoT-enabled sensor nodes. Patients with high glucose content will be prescribed non-glucose-based medicines even if they have a high fever or other medical issues. Edge-AI solutions play a crucial role in such personalized medicine as sensor data is kept close to the end users.

11.2.3 Stress release systems

In the modern world, many people work as machines, while machines have started to work smartly. The stress level of people working in a number of organizations, including academic organizations, remains high due to the committed completion of daily tasks. It is crucial to identify the stress levels of such employees so that employers do not lose these employees and the families of the employees do not suffer hardship.

Edge-AI solutions can solve stress-related issues if the behavior of employees is captured using video-connected systems. The root cause of the increasing stress levels could be sorted out based on the sensed data and inputs from various data points. In this way, the intelligence is applied at the edge, where users are located, without sending the data to cloud services.

The emotions of humans, such as anger, sadness, happiness, disgust, etc, can be easily detected using edge-enabled machines [14]. These devices parse video frames and capture emotions using computationally-intensive deep learning algorithms such as YOLO, CNN, ResNET, etc. Detecting the emotions of school children can pave the way to directly dealing with anomalies and health issues quickly.

11.3 Industrial applications using edge-AI

Edge AI is widely utilized in industry. In general, industries marked several revolutions, such as Industry 1.0, Industry 2.0, Industry 3.0, and Industry 4.0. The processes involved in production environments are enriched with newer techniques and methods, such that the revenue of the industry has remarkably increased. Figure 11.2 provides a high-level overview of the list of industrial

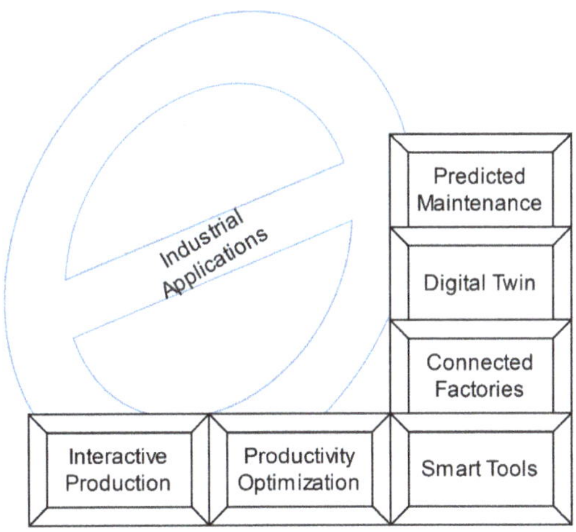

Figure 11.2. Edge-AI-assisted industrial applications.

applications that benefit from using edge-AI. These applications will be discussed in the following section.

In Industry 1.0, the products were developed without any specific process improvement techniques. Product development was slow in the 18th century. During this period, steam engines were used to significantly increase production. With the advent of Industry 2.0, assembly-level production started using electricity. The discovery of electricity in the early 19th century has allowed a massive increase in products over time. However, the assembly production unit has to be carefully managed by a group of managers and full-time workers. Industry 3.0 improved the communication medium between organizations and managers who were involved in production using the internet. The use of internet technologies minimized the time it took to deliver products. For instance, the use of postal communication to order products and subcomponents delayed the production processes. However, email-based communications reduced the time required for end-to-end product delivery to the market.

Industry 4.0 has ushered in a remarkable shift in production processes industry. It has applied IoT-based sensor technologies to automatically detect the requirements of raw materials and decide on the number of products to be delivered on time. The entire process of product development was able to happen without human or semi-human interventions. This led to the application of edge-AI techniques by learning the production requirements and customer requirements in a mass production environment. The edge-AI techniques enhanced the productivity and revenues of industries by modifying the processes involved in the traditional manufacturing system.

11.3.1 Predictive maintenance

Predictive maintenance is an approach where maintenance is supported by IoT-enabled technologies. In earlier cases, maintenance of equipment or machines had to be carried out by engineers after they were worn out or failed. This approach leads to poor productivity in industries, as engineers have to work around failures until maintenance is successful.

In predictive maintenance, sensors will be mounted on machines or factory units that are not frequently reachable by humans. The sensors will collect the details and working conditions of machines periodically. This data is processed to detect machine failures. For instance, the hissing noise of machines may not be noticed by a machine operator who is accustomed to being near to the machine for many years. In contrast, sensors could easily detect such minor variations in the sound produced by machines. Obviously, based on the data and the analytics of the data, maintenance could be carried out in time.

The major advantages of predictive maintenance are low costs, decreased down-time of machines, increased productivity, reduced redundant failures, improved safety, etc. Due to timely maintenance, the lifetime of machines will increase.

Traditional machine learning algorithms such as linear regression, logistic regression, random forests, support vector machines, neural networks, and other commonly utilized algorithms are used in predictive maintenance problems. Owing to structured data input from sensors connected to machines, the prediction algorithms are much faster than a few complicated deep learning algorithms. However, there are use cases to utilize deep learning-based classification algorithms to learn the maintenance-related issues of machines. For instance, ResNet-based learning algorithms could be applied to 3D printing machines to detect nozzle problems or bed leveling issues [17] from the printed outputs of machines.

11.3.2 Digital twin

A digital twin is a visual representation of a real environment's process or product. The term was coined by Michael Grieves in 2002. NASA applied this concept to explore space objects by visually representing them. The idea of the digital twin is to connect both real and virtual objects so that a single virtual instance could manipulate millions of real objects.

The digital twin has multiple distributed sensors to collect real-time data on real objects and process it using virtual objects. For instance, a car manufacturing plant could utilize digital twin virtual objects to mass produce and alter the configurations of tens of hundreds of real manufacturing plants at one time. In this way, the manufacturing processes are productively enhanced using digital twins.

The major components of digital twins are the linking data and actuating connections [15]. The data has to be collected from multiple sensors that are connected to real-world objects. The data could be either centrally processed in the cloud or processed in a decentralized manner using edge-enabled devices. These sorts of predictions or learning inferences require a strong relationship to edge intelligence that should be processed on edge nodes.

The digital twins are deployed using cloud services in the initial stage of development. However, the evolution of edge-based computing systems has increased the application of digital twins in various domains. For instance, manufacturing and controlling car production are remotely controlled using digital twins.

Edge devices are mounted in digital twins to increase the speed involved in the manipulation of physical objects. Otherwise, the data has to be transported to clouds, and decisions have to be taken with a delay in operations. The edge-AI-supported solutions increased the security capabilities of systems. This is due to the fact that the data is not transported to third-party cloud systems. Similarly, the past operations of physical objects or machines could be tracked using edge clusters while delivering intelligence.

11.3.3 Connected factories

Connected factories are cloud-assisted manufacturing platforms that collaboratively share the resources required to produce products. The idea is to provide IoT-enabled computing solutions that could rapidly provision resources from a shared pool of configurable resources via the internet. The connected factories integrate high-end technologies into traditional manufacturing paradigms.

The major advantages of offering edge AI in connected factories include:

1. Increased business: The business operations increase on a distributed cloud-based manufacturing platform. It is because of the shared distribution of raw materials and sub-products that a product is developed. As manufacturing and delivery units are highly distributed, the awareness of the product is high, creating a greater value for the product.
2. Fast production: In cloud-assisted connected factories, manufacturing processes need not wait for hours or days owing to the availability of required resources. Even if the resources are not immediately available, the product manufacturing request could be taken to different geographical locations where resources are readily available.
3. Minimal wastage: There is no need to transport raw materials to a centralized manufacturing plant in connected factories. Due to this approach, the need to order a large number of products, devices, or subcomponents can be avoided. Obviously, wastage due to the development of products is significantly minimized in connected factories.
4. Reduced cost: There are several reasons that reduce the costs involved in manufacturing a collaborative product. Quick production, minimal waste, increased awareness, larger innovations, etc, can directly or indirectly reduce the costs involved in the manufacturing processes.
5. Increased flexibility: It is not necessary that the entire process be centralized in shared factories. The factories and the administration are provided with the utmost autonomy to deliver their sub-products, subject to increasing the objectives. In this regard, the platform offers a huge amount of flexibility in developing collaborative products.
6. Optimization opportunities: The connected factories induce opportunities to optimize objectives. In fact, there are potential trade-offs to designing one holistic solution. For instance, minimizing the costs involved in choosing a factory that has cheaper raw materials could lead to huge transportation costs. Hence, specialized optimization algorithms need to be formulated considering multi-objectives in connected factories.
7. Multi-site production: Another opportunity for promoting connected factories is the possibility of involving multiple production plants from different geographical locations. For instance, a pot could be manufactured in a plant that has sufficient clay nearer to its locations, whereas painting could be carried out in a hot and dry place where natural support is available.

Figure 11.3 showcases the list of advantages that can improve business in connected factories due to the application of Edge-AI.

The purpose of involving edge-AI solutions is to provide predictive maintenance and manufacturing assistance to products [13]. Edge-assisted AI solutions can boost businesses or product delivery situations in such connected factories. Additionally, the feedback of customers could be directly input into decentralized systems, considering the local regions, if edge-AI solutions are developed for connected factories.

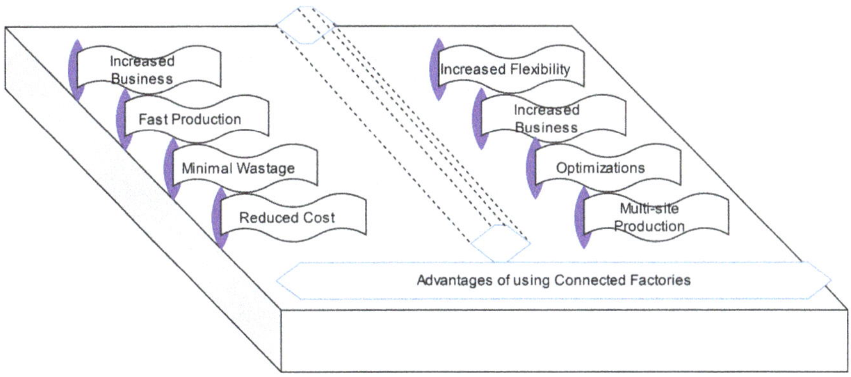

Figure 11.3. Advantages of edge-AI in connected factories.

11.3.4 Smart tools

Manufacturing industries have marked several AI-assisted solutions apart from Digital Twins and predictive maintenance solutions. The tools involving software and hardware combine edge-level AI mechanisms to improve efficiency. A few of the most notable AI-assisted tools and solutions that are vividly utilized in the manufacturing sector are listed as follows:

1. Error detection tools: AI-assisted edge-level solutions have become one of the crucial solutions for detecting errors in machines or industrial processes. For instance, dust intensity mounted on machines could be missed by human eyes. However, edge-AI-based cameras easily detect it and notify users.
2. Automated tools: Several manufacturing units have employed robots or software bots to automatically reason about the functioning of machines. In certain situations, the machines may require further instructions or human intervention. These interactions are avoided by bots. For instance, ordering raw materials, collecting materials from other departments, proactively deciding on the number of required components for production, etc, could be handled by these bots automatically.
3. Night workers: Machinery tools are designed with edge-AI mechanisms to deliver results during the night. For instance, cleaning machines, cleaning workspaces, ordering stationary items or consumables, etc, are carried out by specialized robotic tools during the night. In doing so, the requirement for lights in manufacturing plants can be avoided. Also, the specified tasks, such as cleaning operations, can be undertaken without disturbing existing useful work during the production time of the manufacturing units.

11.3.5 Optimized productivity

Edge-AI solutions are often applied to increase the productivity of industries. They are utilized in two broad approaches: (i) supply chain optimization and (ii) demand prediction. This subsection details these two approaches to increasing the productivity of industries.

11.3.5.1 Supply chain optimization

In general, large-scale manufacturing units will receive tens of thousands of orders from various parts of the globe. With the ongoing use of cloud manufacturing techniques, orders have dramatically increased for a few potential quality manufacturers. Obviously, it becomes difficult to issue commands and deliver duties to subordinates or connected machines manually. Hence, edge-assisted AI solutions will become useful in these situations.

Handling a big-data system in an automatic manner is not a difficult task for any AI-based system. For instance, a large scale computer production unit can keep track of assembled components with appropriate timestamps and delivery maps. In doing so, the AI-assisted software could easily figure out customers who purchased such computers. This approach will be beneficial for replacing the components of computers once they are sold to customers.

The edge-enabled AI systems could be combined with other technologies, such as blockchains. This combination can provide a direct sale deal with customers or a purchase deal with producers. For instance, AI-assisted blockchains can enable the direct purchase of rice from farmers without any dealers or brokers.

Proactive decisions on supply chain management using AI-based assistance can minimize the time to delivery. It can perfectly deliver goods on time, as specified in the pre-assigned rules. This is because the data relating to supply chain operations is assessed and proactively predicted to understand potential hurdles or bottlenecks in operations. The proactiveness of decisions could vary depending on the AI algorithms that are performed in supply chain operations. For instance, federated learning-based edge intelligence can boost the performance involved in supply chain operations due to the collaborative involvement of regional data to make decisions.

11.3.5.2 Demand prediction

The second factor that improves productivity in the manufacturing sector using edge-AI solutions is demand prediction. Many manufacturing plants are unaware of the exact demand for products. For instance, car manufacturers are not aware of potential demand for a particular model in a country due to the requirement of learning a large volume of data. Details such as interest, budget, country inflation rate, previous purchase data, behavior of car owners, etc, could impact the sale of cars.

Edge-AI systems can let producers understand the buying patterns for particular products. This can have direct and indirect impacts on the economy of a country. For instance, the Gross Domestic Product (GDP) rate of countries increases and the inflation rate of countries decreases if the buyer pattern and the usage pattern of products are predicted using AI-assisted systems. Notably, the sale of seasonal products such as fire crackers or Santa Claus outfits increases during November and December every year. This buyer pattern could be predicted using AI-based algorithms that are customized to regional datasets using edge-AI algorithms. North India hardly celebrates Christmas, whereas south India celebrates it.

The authors of [18] designed a demand prediction strategy for manufacturing plants by collecting the real-time demand details of the plant.

11.3.6 Interactive production

Virtual reality-assisted solutions involve edge-AI algorithms for creating interactive production environments in the manufacturing sector. The virtual reality technology contributes to developing manufacturing prototypes in the form of simulations. As the name indicates, virtual reality forms a virtual illusion of the manufacturing platform. If a manufacturer wants to experiment with the modification of the dashboard in a car, virtual reality provides a platform to view it.

Prototyping manufacturing-related products or sub-products of an entire product provides an AI-assisted environment supported by edge devices. The purpose of prototyping manufacturing-related products is to verify and determine the potential hazards or stability features of manufacturing products in a specified environment.

Engineers could experiment with their new designs or prototyping procedures using virtual reality-based solutions. In fact, modifying a window location in a house could have several side effects, i.e., the sunlight might be poor, visibility to specific spots in the environment may be reduced, and sound levels may be detrimentally affected. With the help of edge-assisted virtual reality solutions, the simulation environments could alter the positions of windows, doors, or other lighting conditions in house designs.

11.4 Edge-AI for agriculture

Edge-AI has been a beneficial supportive mechanism in agriculture. Broadly, agriculture-related applications are divided into two broad categories: (a) vegetation-related agriculture and (b) animal-related farming. Vegetation-related agriculture is a category that deals with the application of AI for monitoring crops, soil, plants, and associated accessories, whereas animal-related farming ensures the better living conditions of animals and monitoring them. Figure 11.4 classifies agricultural applications into two broad categories and illustrates the subtopics that will be covered in this section.

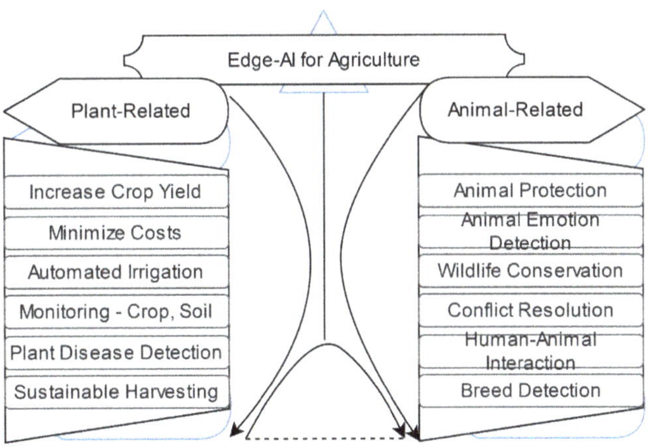

Figure 11.4. Edge-AI-assisted agricultural applications.

11.4.1 Vegetation-related agriculture

Agriculture that attempts to increase crop protection and plant growth falls under vegetation-related agriculture. The reason for involving edge-AI solutions in agriculture started with the need for innovations in the domain. The authors of [16] proposed an agricultural management system that collects information and predicts agricultural benefits based on LoRa technologies: an innovation. The root cause of the need for innovation in major agricultural-related applications is the demand for sustainable food crops in the future.

Edge-AI is utilized in vegetation-related agriculture to enable precision farming and increase the associated yields. There are several reasons that contribute to the necessity of having edge AI, as discussed below:

1. Increase crop yield: AI-based solutions are applied in agriculture to increase crop yields. To do so, it is required to collect appropriate information about growing environments, weather, crop status, etc. Using this information, it is possible to predict productive environments and buyers.
2. Minimize costs: An AI-powered agricultural setup can carefully select pesticides and optimize the requirement of fertilizers, considering the expenditures. For instance, the revenue of apples or mangoes could be increased by accurately detecting the ripeness of the fruit on a daily basis using AI-assisted drones. Similarly, coconuts are harvested using specialized AI-powered drones without manual intervention, and paddy fields are organized to report the daily status of weeds and insects.
3. Automated irrigation: It is not necessary to pump water to a large hectare of land if weather prediction forecasts the possibility of rainfall in a region. An AI-assisted system can automatically drive the pumping system on land so that water is not wasted due to excessive water or near-future rainfalls. Also, the AI-powered solutions could determine the landscape of vegetation lands to detect the required amount of water for irrigation.
4. Crop, soil, and water monitoring: Most of the agricultural solutions combined with AI are supported by appropriate monitoring systems. It is important to monitor the health status of crops, soil, and water in a timely way. In some cases, the water quality is not suitable for vegetation. For instance, laying crops in a high-pH region will not yield as expected. There are several remedies to counteract such variations in the pH value of water. An AI-assisted solution could prevent crops and actuate methods to increase the productivity of agriculture.
5. Disease detection: Buds and smaller plants that are utilized as green vegetables are prone to diseases. Almost all trees require special attention on a daily or weekly basis to protect them from serious diseases or insects. In fact, it becomes difficult for owners or farmers to keep track of these different categories of diseases. For instance, coconut trees encounter several types of disease in the southern part of India. They are dry leaves, yellow leaves, insect bites, and caterpillar bites. These diseases could be automatically detected and forecast so that appropriate treatments are possible.

6. Sustainable harvesting: Sustainable harvesting is a management process in which agriculture is handled in a natural way. For instance, utilizing rainwater, wind energy or solar-powered pumping stations, natural manure utilization, etc. In [1], the authors have studied the wider components, such as drones, for enabling sustainable harvesting techniques in agriculture.

11.4.2 Animal-related farming

Agriculture that improves the living conditions of animals, pets, and wildlife is categorized as animal-related farming. Edge-AI techniques could be applied in this domain in multiple ways. The most commonly utilized applications are described below.

11.4.2.1 Animal protection

Edge-AI techniques remain pivotal in protecting animals from danger. Wearable devices could be mounted on animals to constantly monitor their activities and detect health-related issues. Accordingly, veterinary doctors could be deployed to targeted locations to provide medicines and vitamins. Similarly, animals could be automatically warned or prevented from attacks using edge-enabled IoT solutions. For instance, elephants could be chased away from attacking other animals by automatically producing a buzz of bee swarms.

11.4.2.2 Animal emotion detection

Similar to human emotions, animals do express emotions in various ways. Identifying emotions such as happiness, sadness, anger, etc, could maintain animal welfare and keep them mentally healthy in the wild and in homes. It is crucial to detect the emotions of domestic animals as they live alongside humans.

Recently, researchers have studied the emotions of elephants. There have been efforts in the past to study the emotions of other animals, such as cats, dogs, pigs, and others. Emotion detection in animals is a subfield of affective computing in AI. The main idea behind identifying emotions is to study the state of animals using sophisticated deep learning techniques such as CNNs. In fact, several subparameters, such as heart rate, brain activity, facial expressions, and body gestures, can provide valuable information about the mental state of animals.

There are a few deep learning algorithms that can detect the emotions of animals. For instance, algorithms such as CNNs, ResNet, lightweight CNNs, and YOLOs can predict the emotions of animals based on pre-trained or custom-trained models. The algorithms have increased prediction accuracy if multi-modal information is fed into them. Notably, detecting the emotions of elephants using facial expressions and trumpeting noises can easily increase the prediction accuracy while detecting the emotions.

Animals exhibit emotions such as happiness, sadness, anger, fear, etc, [7, 8]. The detection of animal emotions includes psychology, biology, and veterinary medicine. Traditionally, emotions were manually detected by trained professionals or animal owners [11]. By suggesting remedial actions based on their emotions, the animals can live peaceful and happy lives in the wild or in homes. Also, the medication of animals could be monitored by assessing their emotions while treating them.

Detecting the emotions of animals is broadly classified based on the detection techniques as follows:

- Invasive method: Invasive methods rely primarily on wearable sensors with normal biometric data and physiological indicators that are collected in real-time. This approach deals with individual animals. The major disadvantages of this approach includes handling a large volume of sensor data relating to emotions and mapping it with the characteristics of animals in real-time. Also, it is a challenging task to detect the emotions of a group of animals, as wearable sensors are often mounted on individual animals that are mobile in nature.
- Non-invasive method: In contrast, non-invasive methods are based on collecting sensor data from facial expressions, animal behaviors, and body gestures. Among them, the facial expressions of animals are broadly utilized to study the emotions of animals. In the non-invasive method, sensors are mounted on nearby walls, trees, or room shelves to capture the gestures of animals. As the sensors are immobile, they can provide stable data for detecting emotions. The most commonly utilized emotions in animals include happiness, sadness, fear, disgust, neutral, etc. As mentioned earlier, the accuracy of predictions increases by adding multi-modal input data such as images, texts, voice, and video frames.

11.4.2.3 Wildlife conservation

Edge-AI systems can contribute to conserving wildlife species. In general, there are several programs and organizations involved in protecting forests or wild animals. For instance, tourist centers have adopted policies to provide a peaceful environment for wild animals, avoiding disturbing them throughout the day. There have been programs to organize sponsorship activities while animals are in distress. Also, organizations have become stricter in preventing forests becoming polluted by plastic waste from humans, which can detrimentally affect specific creatures.

A robust surveillance system could be adopted in forests to protect animals with the utilization of edge AI solutions. Animals could be rehabilitated in newer locations or forests if they are diligently monitored in real-time. In some situations, animals have to be relocated due to the scarcity of water or food materials, or they have to be provided with appropriate food by notifying the relevant forest officials. All these activities increase wildlife conservation with the assistance of edge-AI systems.

At times, especially during natural disasters such as floods or fire incidents, the animals living in forests have to be protected by assessing the remote locations and ensuring the requirement of assistive materials. Similarly, forest officials have to be protected from the wild animals that they are protecting. Hence, edge-enabled systems could become a solution to preserve wild animals and balance their livelihoods in contradictory environments.

11.4.2.4 Human–animal conflict resolution

Human–animal conflicts are not new issue. They have been in existence for centuries. However, the traditional approach of resolving such conflicts by killing animals has reduced the animal populations, especially a few endangered ones. This is because humans encroach on animals' living spaces by increasing human populations.

Human–animal conflicts are very prominent in tropical countries such as India, Kenya, Costa Rica, etc. For instance, elephants attack humans at night. This is because people are gradually encroaching on forest locations to plant vegetation. Similarly, there are lion attacks in Kenya as they are often visited by wildlife enthusiasts, and there have been several news reports highlighting the attacks of bears on humans or police guards in Canada. Many of these incidents could be prevented by a specialized solution that could automatically notify humans about the dangers.

Traditionally, there have been approaches to providing electric fencing to prevent elephants from intruding on the farms of villagers. However, such fences have become a burden to several other endangered species living in the same region. Solutions such as edge-AI mechanisms that could detect the arrival of elephants and enable electric fences could be a solution to the problem.

11.4.2.5 Human–animal interaction

In practice, humans can communicate with each other using conversations that are learned in primary or secondary schools. However, animals do not have such structured learning systems. Obviously, there is a need for smooth interactions between humans and animals for several reasons, as listed below:

1. Work Sharing: Many animals work alongside humans. For instance, dogs protect humans and hunt other animals; elephants carry logs based on the instructions provided by mahouts; donkeys are utilized to carry heavy loads, etc.
2. Emotion Sharing: Animals are a stress-relievers for some humans. Many people spend a considerable amount of time interacting with their pets, such as cats, dogs, parrots, etc. The reactions and emotional intelligence shared between them have several advantages.
3. Therapies: Animals are used to solve some medical issues that afflict humans. For instance, animals such as cats, dogs, horses, and so forth can provide therapy to humans when they are facing stress. A few animals play and birds sing to improve the mood and well-being of humans.

Edge-AI systems could enrich the interactions and relationships between humans and animals. It could also prevent conflicts. For instance, if elephants were hungry, they could cause physical damage to humans. Such potential problems could be identified, analyzed, and notified using edge-AI-based novel solutions [6].

11.4.2.6 Animal breed detection

Animal breed detection is a prominent example that utilizes edge-AI systems. It remains tougher for humans to exactly determine breeds and their characteristics. With the application of edge-AI-assisted technologies, videos could capture information and engineer features from the dataset to classify breeds with ease. For instance, the authors of [10] designed a methodology using edge AI to classify breeds among cows. The approach could be helpful to veterinary doctors or concerned authorities.

11.4.3 Use-case: elephant emotion detection framework using deep learning

Recently, we have designed a framework that predicts elephants' emotions in real-time. The framework was integrated into a mobile edge device, e.g., a smartphone, to help people who are not animal experts gain access to elephants. The idea of performing elephant emotion detection is to increase the working capability of elephants by understanding their moods. The framework utilized labeled datasets of elephant emotions. Later, object detection and emotion detection was performed.

11.4.3.1 Traditional approach

Traditionally, object detection tasks consist of three subtasks:

1. Region selection: In the past, regions from parsed video frames were utilized to detect appropriate regions where elephants are located. It resembles template matching and sliding windows with different aspect ratios and sizes for selecting those regions. In fact, this process required a large computational workforce and time.
2. Feature extraction: Often, researchers used histograms or Haar classifiers to extract features from images. It was also practiced in the past to avoid potential issues such as occlusion, illumination, background noises, etc while extracting features from images.
3. Classification: Traditionally, SVM or similar types of classifiers have been utilized to classify objects from images. Although several research works interested the market, the traditional classifiers kept simple processing steps for several applications and researchers.

11.4.3.2 Deep learning approach

The application of deep learning models for object detection and tracking has marked a unique scalable factor in innovations. It did not restrict the training to certain pre-defined manual features. In contrast, deep learning had the most robust representation in automating the feature engineering processes of data. The deep learning approach increased its cover by developing several learning algorithms that are categorized into two-stage detection and one-stage detection stages. Algorithms such as CNNs correspond to two-stage detections, and YOLO corresponds to one-stage detectors.

Unlike two-stage object detectors, the skipping of the region proposal process results in a shortening of inference time, which is valuable when performing real-time prediction. The relatively satisfied accuracy and low latency are reasons that YOLO is suitable for development as a mobile application. YOLOv5 inherited the strength of YOLO and made improvements with precision and the ability to detect multi-scale objects. A detailed description of different deep learning algorithms is discussed in [2].

11.4.3.3 Edge-AI lightweight approach

Edge-level intelligence in detecting the emotions of elephants is carried out using IoT and similar high-end technologies. The IoT technology combined with deep learning grows rapidly to automate predictions. It takes advantage of bringing deep learning applications closer to consumers. This real-time prediction is likely to be realized as a reduction in latency caused by data transmission between the edge and cloud.

In the elephant detection framework, containers are utilized to provide a lightweight solution. Containers share the host operating system kernel and only include the libraries and dependencies needed to run the application. This results in a smaller footprint and faster startup time. As an OS-virtualized software platform, a self-contained container resolves the dependency issue with the respective OS. Thus, it is easy to create, deploy, and run applications using tools such as Docker Compose. Docker avoids performance and cost overhead and ultimately offers better performance. It is able to manage the virtualization process and can, therefore, be used to distribute, deploy, and manage cloud and edge applications assigned to the clusters.

11.4.3.4 Elephant emotion feature detection

The emotion detection approach consisted of two pre-trained YOLOv5 models: (a) The first was the YOLOv5 model trained on the COCO dataset, which is used to detect the elephant. (b) The second was the YOLOv5 model that was trained with an elephant dataset with seven emotional features labeled: ‘happy eye’, ‘happy nose’, ‘happy mouth’, ‘sad eye’, ‘sad nose’, ‘angry eye’, and ‘angry ear’.

Once the model was created, inferring the emotions of elephants was detected by following the steps: (i) parsing video inputs, (ii) chopping them into frames, (iii) detecting elephants from frames, (iv) identifying bounding boxes and their positions, (v) classifying emotions with scores, and (vi)listing the scores and concluding the emotions. The entire framework involves mobile phones or edge-enabled devices to perform the intelligence on edge-level nodes.

11.4.3.5 Metrics utilized

To evaluate the algorithms associated with elephant emotion detection, we utilized the following metrics: (i) precision (equation (11.1)), (ii) recall (equation (11.2)), and mean average precision (mAP) (equation (11.4)).

$$\text{precision} = \frac{TF}{TP + FP} \tag{11.1}$$

$$\text{recall} = \frac{TP}{TP + FN} \tag{11.2}$$

$$\text{accuracy} = \frac{TP + TN}{TP + FN + FP + TN} \tag{11.3}$$

$$\text{mAP} = \frac{\sum_{i=1}^{N} AP_{\mathrm{i}}}{N} \tag{11.4}$$

where true positive (TP) means the cases in which the prediction was yes and the actual output was also yes; true negative (TN) means the cases in which the prediction was no and the actual output was also no; false positive (FP) means the cases in which the prediction was yes and the actual output was also no; false negative (FN) means the cases in which the prediction with NP and the actual output were also yes; AP means average precision; N means number of classes; TP means number of true positives; FP means number of false positives; FN means number of false negatives.

11.4.3.6 YOLO-based implementation details

To implement and capture elephant emotions, cameras were utilized in real-time. Additionally, as elephants are rarely found in many geographical locations, the videos available on YouTube links were considered for predicting emotions. In the work, a Pytorch framework was utilized to develop deep learning algorithms. The captured videos, either real-time screening videos or offline videos, were parsed and sniffed using `OpenCVPython` packages. The entire framework was integrated into a mobile phone, a type of edge device.

In the work, YOLOv5 was utilized as a single-shot detector to study the emotions of elephants. Initially, it is required to create a dataset that captures the emotions of elephants. The datasets are prepared such that the emotions are annotated using labels. In the work described above, we utilized `labelImg` software to create labels. Using the software, we could create bounding boxes to label them as distinguished classes. Here, to detect the emotions of elephants, the faces of elephants were labeled as happy, sad, fearful, and angry.

As we are familiar with the deep learning basics, it is a known fact that a training phase is initially required using algorithms to study the emotions of elephants. To do so, Python packages are imported into the implementation.

```
import torch
import torch.distributed as dist
import torch.nn as nn
```

Similarly, the YOLO model was imported to train the elephant's emotions, as follows: from models.yolo import Model

```
from models.yolo import Model
```

After setting the required configurational parameters and paths where the trained model has to be saved, the elephant emotions are trained for a specified number of epochs, as follows:

```
for epoch in range(start_epoch, epochs):
 callbacks.run('on_train_epoch_start')
 model.train()
```

The implementation was parallelized using GPU-assisted CUDA programming, and the training models were saved at any point in time with specified states in checkpoints so that they could be restarted from any point in the training stage. In a Python-based implementation, the checkpoints and YOLO models are saved as follows:

```
torch.save(checkpoint, state)
```

Once the training is complete, the trained model is stored in a folder so that it can be utilized for further testing or validation purposes. The training and testing procedures are scripted such that commands are sufficient to train and test datasets. For instance, the training and testing features are activated using the following commands:

```
Training: python train.py -data data/elephant.yaml
 -cfg models/yolov5s.yaml -batch-size 64
 -name Model -epochs 2000
Testing: python detect.py -source Elephant -weights
Model.pt
```

It can be observed that the training activity is carried out for elephant images that are labeled datasets. The training phase required the name of the model and the number of iterations for developing the model.

Later, the model was converted to a lightweight TFLite model so that it could be ported to small edge-enabled devices for further inferences. In the testing phase, the inferences are carried out in real-time by capturing the emotions of elephants using mobile devices.

11.5 Edge-AI for forensics

Forensics is a science that investigates crime situations using technologies and mechanisms. Traditionally, manual procedures have been adopted to detect crime-related investigations using forensic scientists. However, with the advent of AI and edge intelligence, forensics became a solid manifestation to identify criminals or crime scenarios [12]. The technologies assisted forensic scientists to detect situations with ease.

11.5.1 Forensic applications

Although forensics is important, it is better to assess possible situations in which artificial intelligence can play a vital role. Toward this end, a few lists of forensic applications are analyzed as follows:

1. Site analysis: Detecting crime sites and investigating the probability of crime situations can be carried out with the assistance of modern IoT-enabled technologies. Evidence such as body tissues, hair, blood, etc, that relates to biological factors can be tracked using deep learning-assisted solutions. Similarly, evidence relating to tire pressure or the footprints of criminals could not be easily tracked by human eyes if potential variations were performed. It is crucial to detect the fingerprints and blood stains of criminals. IoT-enabled solutions or virtual reality-based metaverse solutions could be developed to accurately detect with explainability using edge-assisted AI solutions.
2. Digital forensics: Digital forensics is a subfield of science that requires intelligence to learn from crime-related data. Additionally, it has to deal with the approach to storing and preserving data. In fact, data could be in several formats, such as texts, voice, video, or images. The application of sensor-enabled technologies at edge-level nodes enables faster investigation of cybercrimes. To preserve data and findings in forensics, blockchain-assisted AI technologies are utilized in digital forensics.
3. Emotion/lie detection: Facial expressions could be tracked and replayed with sufficient evidence while applying edge-AI solutions. In general, the human brain has two main parts: the left hemisphere and the right hemisphere. The left hemisphere is used for analytics, and the right hemisphere is responsible for creativity. A person expressing the truth of a crime scenario turns their eyes to the left to collect information from the left hemisphere of the brain, whereas a person attempting to lie due to criminal offenses will more often move their eyes in the right direction. This means that the person wishes to create a new story as part of the creative situation. The lie might not be correctly detected by police investigators if they are not skilled. However, edge-AI solutions can easily capture such factors with ease and with explainable evidence.
4. Handwriting/script analysis: Suicide often has specific evidence from the victims in terms of notes. Suicidal notes or scripts mentioning the cause of death by a person before the death are considered crucial evidence for forensic scientists. However, the task becomes extremely arduous for them to correctly analyze the writings owing to various factors; for instance, a person in fear or stress may not write properly. Hence, edge-AI-assisted algorithms are required to correctly analyze the handwritten scripts of suicidal notes or death notes.

11.5.2 Edge-AI requirements

Although there are several edge-AI advantages, the platform that supports forensics should be designed to protect data or evidence from potential defaulters.

The learning models or prediction engines that handle data should not transfer the data to third-party providers, including cloud servers. Hence, edge-AI solutions are convincingly accepted by a wide sector of researchers and practitioners.

The edge devices that are involved in analysing the evidence from crime sites have to be well protected from security attacks or malware conditions. If not, there are chances of levying wrong decisions that could lead to incorrect conviction and imprisonment. It not only affects the psychological improvements of individuals but also the families belonging to victims.

Typically, edge-AI solutions can provide wrong inferences or evidence. The explainability of reaching such decisions could have flaws or inaccuracies. Obviously, the outcome could lead to failure. In order to prevent future implications, all evidence and inferences need to be well documented and preserved in specialized data storage units with robust technologies such as blockchains. The blockchain technology has an approach to preventing data tampering situations due to the offer of providing a trustless environment using a network of computing nodes.

The edge-AI systems have to integrate well with the traditional forensic mechanisms. For instance, the input from forensic scientists has to be integrated into existing training models. Also, models have to fit within the memory size of utilities or appliances that are utilized by forensic scientists. The edge-AI systems that are designed for forensics will have to support as many communication protocols as possible to deliver a sufficient quantity of services.

The services involving cloud providers have to be supportive of delivering appropriate information to scientists or users. There have been efforts in the past to study cloud-specific forensic information. Researchers have developed tools and libraries that could provide the necessary information from disks of virtual information to analyze crime sites or data. Similarly, there are requirements for edge-AI systems to capture cloud-specific data that is deleted or moved to other repositories. Such data could be a valuable source for providing evidence.

11.5.3 Edge-AI forensics: stages

Applying edge AI to forensics is carried out in various stages, especially when data are involved to process and collect evidence from crime sites. Typically, a more generic approach to performing forensics using edge-AI systems is listed as follows:

- Resource identification: The first stage to performing forensics using edge-AI-supported platforms is to define the amount of resources that need to be involved. A few resources may be required to assess cloud-specific devices, some other resources may be utilized to assess edge-specific networking devices, some sensor-attached resources could be engaged in collecting data for the analysis, etc. Hence, it is important to understand the most possible resources that might be required in detecting crime sites.
- Data collection: The next step is to collect the required data to analyze evidence using sensors or similar technologies nearer to the crime site. The sensors are connected to edge-enabled devices for faster processing of data

and filtration of irrelevant data. Additionally, data that is associated with cloud services or third-party providers needs to be collected with sufficient indexes and storage manifestations for future analysis.

- Data storage: After establishing sufficient evidence, the relevant data and details have to be well documented and stored in reusable storage units. The documentation, mostly, has to be written in legal terms and court-related law-firm terms so that lawyers and prosecutors will be in a position to express the edge-AI-supported evidence appropriately. Special attention has to be given to instantiating high security and replications for the data. Also, the tenure of this data storage has to be longer than 15 to 30 years owing to the diversion in cases even after the judgment of the courts.
- Evidence delivery: The last stage of the edge-AI-supported forensic system is to deliver the findings in a legal format. Presenting the evidence has to be user-friendly for practitioners and forensic scientists. If possible, the delivery of evidence has to be represented at various levels, marking the complexity involved in the explainability of crime incidences. For instance, the public does not want to know the exact reason for specific evidence of crimes, whereas scientists will require detailed information about the reasons for such evidence and the countermeasures to prevent them in the future. The delivery of such evidence should spark future innovations that could be applied in the system as part of the learning experience of handling crime.

11.6 Edge-AI for mobility/logistics

AI-assisted solutions have been beneficial in the transportation sector while performing mobility or logistics in busy cities. For instance, traffic management, smart vehicular communications, autonomous vehicles, logistics planning, shared mobility, distributed product launches, etc, have been enriched with intelligence or learning assistance in modern products or solutions [9].

The importance of edge-AI solutions in the transportation sector is to deliver agile reasoning and innovative productization. This section details the application of edge AI in four broad applications in the transportation sector.

11.6.1 Smart mobility

Smart mobility has remained an important ingredient in smart cities. India has aimed to strengthen smart cities in 100 different regions. The major difference between a normal city and a smart city is the provision of vast support features that assist city residents. Smart mobility in smart cities has different application dimensions:

1. Traffic flow management: Edge AI-supported applications can automatically drive traffic management in busy streets and city centers. The sensors need to be mounted on traffic corners, and edge device-supported intelligent algorithms can perform intelligence while controlling the traffic of vehicles. Intelligence can automate the traffic flow even when emergency vehicles need to travel during busy hours on the streets.

2. Parking indicator: In many cities, there are several parking areas that are not seen by drivers while driving their vehicles. The edge-AI-supported parking lot identification system can detect empty lots for parking vehicles, considering the time to reach the parking destination and city preferences. This not only favors drivers but also the cities, preventing a large number of vehicles roaming around the busy streets searching for parking lots.
3. Noise control: Smart mobility can ensure effective communication between vehicles using robust communication mediums. Recently, the enhanced utilization of 5G networks paved the way for vehicles to communicate with each other in the blind spots of cities or streets. This notion of vehicle-to-vehicle communications automatically controls the speed of vehicles and prevents potential accidents without the presence of edge-AI-based solutions. The practice of pressing horns all the way down to the street can be minimized to a negligible level if edge-AI-assisted solutions are regularly provided in all parts of the cities or countries.

11.6.2 Shared mobility intelligence

COVID-19 provided many innovations in the transportation sector, as COVID-19 patients were isolated from public transport. In fact, every country promotes shared mobility, where multiple passengers belonging to various communities and regions travel in a single vehicle such as a bus, taxi, train, or airplane. The countries gain economic improvements by motivating their citizens to utilize public transport with limited convenience compared to personal cars.

There were many challenges in the traditional shared mobility principles: (i) it was not possible to keep track of isolating patients in shared vehicles, (ii) collaboration between different departments, such as the justice department, police department, social security services, hospitals, etc, was not inherently practiced due to the non-availability of technologies, (iii) communication between vehicles and processing capabilities of devices were not sufficient to realistically handle situations, (iv) inefficient utilization of shared vehicles, etc.

The shared mobility assisted by edge AI and blockchains has recently marked a traction in the transportation sector. For instance, recently, a blockchain-enabled shared mobility (BESM) architecture was proposed to enhance the features of the shared mobility concept [4]. The synopsis of the architecture is to suggest that residents travel in a shareable vehicle, considering the air pollution status and COVID-19 status of regions. The idea behind performing such seat allocation is to increase the utility of shareable vehicles with more intelligence, taking into consideration the health status of humans.

To do so, an edge-AI system should capture the details of available shareable vehicles and the demand to utilize such vehicles at a particular time in a city or street. Additionally, a blockchain-based ledger needs to be started with a genesis block so that the transactions of seat allocation are recorded in the blockchain ledgers. To record the information in the ledger, the latitude and longitude of the traveler's location, travel destination location, health information, current air quality information of travel destinations, etc, are collected and processed in the blockchain. Obviously, the information will be iteratively shared among all travel spots to regionally calculate the number of available seats in the shareable vehicle. Thus, several parameters with connected participants, including transport authorities and hospitals, are involved in deciding the seat allocation of a shareable vehicle to travelers. More detailed information about the architecture that can include edge-AI solutions or algorithms for allocating seats in a shareable vehicle is provided in [4].

11.6.3 Logistics planning

Automating services in smart cities is one of the primary tasks for city authorities. Planning logistics and automatically delivering services are part of the automation processes of smart cities. Edge-level intelligence has played a crucial role in performing automation processes by monitoring and analyzing data. One of the major objectives is to improve resource utilization and optimize logistics-related travel in cities.

Manhole detection and logistics planning are two of the applications under logistics planning. In fact, manholes are subject to failure, especially during the rainy season. There is a high chance that the level of manholes will increase, leading to drainage overflow. Or, manual monitoring and logistics planning to pick up full pipelines is a difficult task. Hence, in the past, work was undertaken to study the application of optimization and intelligence while planning the logistics for repairing manholes.

The manhole detection and logistic planning system consisted of sensors that monitored the depth of manholes. For this, ultrasonic sensors that detect the distance between obstacles are mounted below the manhole-closing lid. The sensor data is collected and processed using edge devices. The devices automatically parse data and preprocess it so that only relevant data is submitted to cloud or hierarchical computing nodes.

In the system, there were three services that benefited from the assistance of cloud-based resources. The services are listed as follows:

- Manhole level indicator service: this service collects manhole levels from edge nodes that are connected to level-indicating sensors.
- Logistics planner service: the purpose of this service is to collect transport details and the available resources, such as transportation vehicles, manpower (if any), etc.
- Scheduler service: this service has codes that can schedule resources based on smart city information and manhole levels using scheduling algorithms. In the work, the threshold acceptance algorithm was utilized to plan the schedule to perform logistics in cities.

11.6.4 Product placement

Smart mobility is not just for vehicles or associated services. It can also play a pivotal role in indoor transportation. For instance, shopping malls are increasing day by day in cities, and small shops are being put out of business by these shopping giants. People prefer to purchase from such malls as they get a reduced price for products, and they are entertained an entire day while visiting such malls.

Transportation and logistic planning in huge malls within indoor environments could be a challenging aspect as several constraints are involved in the processes. Notably, the shops are interested in gaining business and attracting customers throughout the period of operations. For instance, placing products in the empty slots of showcases in malls is a mandatory step to attract customers and increase the revenue of businesses. This process requires a strong bond between edge-AI and mall-equipped devices.

In the past, an in-store product placement approach was developed using LiDAR sensors [5]. The idea behind this approach was to quickly identify the empty slots in showcases in supermarkets and replace them with customer preferences. In fact, the idea fits very well in huge shopping malls where shop employees cannot keep track of empty products or the preferences of customers. Also, the approach could be beneficial when the number of customers increases, especially during Christmas or Diwali holidays.

11.7 Conclusion

There are tens of thousands of edge-AI applications on the market. They are developed in various domains such as healthcare, industry, agriculture, transportation, digital forensics, etc. This chapter describes the potential edge-AI applications that are being experimented with by researchers or practitioners. The chapter showcases an approach involving edge-based mobile phones to detect the emotions of elephants. Additionally, a list of novel ideas has been shown in this chapter so that readers can explore the application of such ideas using frameworks or orchestration platforms discussed in the previous chapters of this book.

References

[1] AlZubi A A and Galyna K 2023 Artificial intelligence and internet of things for sustainable farming and smart agriculture *IEEE Access* **11** 78686–92

[2] Benedict S 2022 Deep learning technologies for social impact (https://iopscience.iop.org/book/mono/978-0-7503-4024-3.pdf) (accessed 17 February 2024)

[3] Benedict S 2022 IoT-enabled remote monitoring techniques for healthcare applications—an overview *Inform. J.* **46** 131–49

[4] Benedict S 2022 Shared mobility intelligence using permissioned blockchains for smart cities *New Gener. Comput.* **40** 1009–27

[5] Benedict S, Lakshin Kumar R and Baranidaran M 2023 In-store product placement using lidar-assisted discrete PSO algorithm *Information, Communication and Computing Technology* ed J Abawajy, J Tavares, L Kharb, D Chahal and A B Nassif (Cham: Springer) pp 74–87

[6] Chiu Y-J, Benedict S and Gerndt M 2023 E2d2: elephant emotion and distraction detection framework using edge-enabled yolov5 deep learning algorithm *2023 IEEE International Conference on Contemporary Computing and Communications (InC4)* pp 1–6

[7] Feighelstein M G 2022 Towards automatic recognition of emotional states of animals *Proceedings of the Eight International Conference on Animal-Computer Interaction ACI '21* (New York, NY: Association for Computing Machinery)

[8] Ferres K, Schloesser T and Gloor P A 2022 Predicting dog emotions based on posture analysis using DeepLabCut *Future Internet* **14** 97

[9] Gong T, Zhu L, Yu F R and Tang T 2023 Edge intelligence in intelligent transportation systems: a survey *IEEE Trans. Intell. Transp. Syst.* **24** 8919–44

[10] Gupta H, Jindal P, Verma O P, Arya R K, Ateya A A, Soliman N F and Mohan V 2022 Computer vision-based approach for automatic detection of dairy cow breed *Electronics* **11** 3791

[11] Neethirajan S, Reimert I and Kemp B 2021 Measuring farm animal emotions—sensor-based approaches *Sensors* **21** 553

[12] Razaque A, Aloqaily M, Almiani M, Jararweh Y and Srivastava G 2021 Efficient and reliable forensics using intelligent edge computing *Future Gener. Comput. Syst.* **118** 230–9

[13] Resende C, Folgado D, Oliveira J, Franco B, Moreira W, Oliveira-Jr A, Cavaleiro A and Carvalho R 2021 Tip4.0: industrial internet of things platform for predictive maintenance *Sensors* **21** 4676

[14] Rincon J A, Julian V and Carrascosa C 2020 Towards the edge intelligence: robot assistant for the detection and classification of human emotions *Highlights in Practical Applications of Agents, Multi-Agent Systems, and Trust-worthiness. The PAAMS Collection* ed F De La Prieta, P Mathieu, J A R Arango, A El Bolock, E Del Val, J Jordán Prunera, J Carneiro, R Fuentes, F Lopes and V Julian (Cham: Springer) pp 31–41

[15] Savaglio C, Barbuto V, Awan F M, Minerva R, Crespi N and Fortino G 2024 Opportunistic digital twin: an edge intelligence enabler for smart city *ACM Trans. Sen. Netw.* https://doi.org/10.1145/3616014

[16] Sharofidinov F, Muthanna M S A, Pham V D, Khakimov A, Muthanna A and Samouylov K 2020 Agriculture management based on lora edge computing system *Distributed Computer and Communication Networks* ed V M Vishnevskiy, K E Samouylov and D V Kozyrev (Cham: Springer) pp 113–25

[17] Sunny B C, Benedict S and Rajan M P 2023 Electrical energy estimation of 3d printing jobs for industrial internet of things (IIoT) applications *Rapid Prototyping J.* **29** 1592–603

[18] Wang W, Zhang Y, Gu J and Wang J 2022 A proactive manufacturing resources assignment method based on production performance prediction for the smart factory *IEEE Trans. Ind. Inform.* **18** 46–55

IOP Publishing

Edge Intelligence
Deep learning-enabled edge computing
Shajulin Benedict

Chapter 12

Business opportunities using edge-AI

In the modern world, digital businesses have increased, especially in the post-COVID-19 era. Business operations rely on robust technologies such as IoT, AI, and robotics to automate daily processes to increase the turnover of companies. In this chapter, we understand the advantages of applying edge-AI techniques to digital businesses. Toward this end, the purpose of digital businesses, cost-improvement techniques using edge-AI and serverless platforms, and the other few challenges are discussed.

12.1 Digital business

Digital business is the process of applying digital technologies to improve business models. This includes approaches to attract tens of thousands of customers, to increase the smartness of businesses, to offer trust while performing digital communications, etc.

Digital business has increased revenues for several organizations, including retail businesses and educational sectors. The application of cloud-based services reduced the need for purchasing high-end machines or servers. The clouds minimized the requirement for manpower to prevent machines. Integrating social IoT and semantic IoT tailors the business to a particular person or organization [2].

The inclusion of digital business in traditional business operations in sectors such as agriculture, healthcare, education, etc, can drastically increase the visibility of businesses. For instance, automatically recording the health status of patients and deciding on suggested solutions can increase patient's satisfaction and doctor's service ability.

The application of digital business can save time in services or business operations. For instance, efficient supply chain management uses logistics and digitally supported platforms to increase the deliverables to various geographical locations on time. Obviously, the time required for the process is minimized.

doi:10.1088/978-0-7503-5593-3ch12

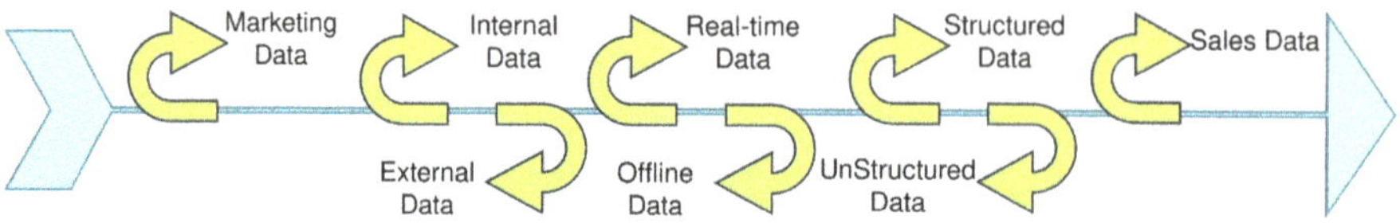

Figure 12.1. Business operational data.

Subsequently, the manpower and the other associated resources could be optimized for several parallel actions.

Data is a crucial source in excelling business operations. The data involved in business operations falls into several categories (see figure 12.1), as listed below:

1. Marketing data: Marketing data is the data that is collected either using machines or human groups to inform marketing campaigns for products. The marketing data involves advertisements and their associated results. The results are automatically evaluated based on the advertisement results that are carried out by individual machines. Otherwise, the results are showcased through individual marketing personnel in any specified geographical location or region. Marketing data remains a key to business operations because the return on investment can be calculated based on the marketing values.
2. Sales data: This data is necessary in business operations to understand selling patterns and volume of sales. The selling pattern depends on national holidays, events, the wealth of people in a society, etc. The pattern can determine the number of sales that will happen in future years or months. Similarly, it can determine the communities that are interested in purchasing the products.
3. Internal data: Data that is collected and utilized within an organization is termed internal data. This data provides a large amount of information about the growth of an organization, including employees' growth and well-being. This data can improve business opportunities as the internal information contains many contents and suggestions. In contrast, external data are data that correspond to an organization's external image or perspective. For instance, the external data of an educational institution could be the family details or judicial records of students that are collected outside the organizational venue.
4. Real-time/offline data: In the majority of businesses, real-time data is utilized. In real-time, the data is collected while the application is running on compute nodes. The real-time data can be influential in making timely business decisions. Obviously, it is not suitable to opt for offline data. However, it is expected in some applications that the amount of business data could be larger. This large volume of data cannot be processed in real-time due to decision overheads. Hence, offline business data could be useful in some applications.

5. Structured/unstructured/semi-structured data: Data that is utilized in performing business operations could be structured, unstructured, or semi-structured in nature. The structured data will have pre-defined parameters and data fields that could be utilized by prediction algorithms such as Random Forests, Support Vector Machines, or XGBoost. In some cases, the data is incomplete or cannot be completely categorized. Hence, a specific methodology needs to be followed to process semi-structured data to increase business operations. However, when the data is large enough, it is nearly impossible to develop or categorize features from the data. In such cases, the data are unstructured: i.e., they are not categorized by pre-defined features. Accordingly, a few deep learning or unsupervised learning algorithms are required to handle such unstructured data to boost businesses.

Additionally, in business operations, the transactions could be automated and recorded in a digital ledger. Transactions, in the past, were carried out on a one-to-one basis. Although a strong trust is maintained in this type of business, the scalability of the business is minimized. In later years, the establishment of trusted parties such as banks and central control authorities increased transactional capabilities. For instance, a person from India could make transactions and undertake business with a person in another country without them knowing each other. In this approach, banks or third parties provide trust while performing business. However, the failure between the third-party organizations and the seller or buyer introduced novelty in digitalization. Notably, researchers introduced the concept of blockchains, where a trustful network provides trust to people making transactions.

In summary, the realm of the business market has increased with the advent of digitalization and allied edge-enabled solutions. As a next level of innovation, the application of edge-AI in business operations can significantly increase revenue or service abilities. Similarly, there are approaches to improving the costs involved in advancing technologies, such as serverless function-based architectures (see figure 12.2). This chapter explains the possibilities of enriching business operations either by imparting digitalization or by minimizing costs involved in the technological developments of edge-AI platforms or frameworks.

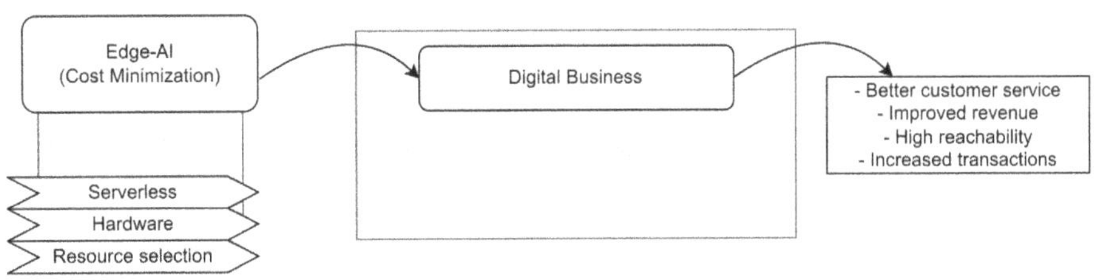

Figure 12.2. Cost improvements: infrastructure and digitalization.

12.2 Economic impacting factors

Edge computing relies on memory and computing requirements. Configuring memory for edge-AI applications and selecting appropriate resources for such tasks can improve the costs involved in applications [10]. Also, the deployment procedures adopted in such edge-AI applications will have impacts on the associated costs. This section explores the potential cost-improvement factors that could be adopted in edge-AI frameworks or platforms to improve the memory, compute, and deployment policies of edge-AI applications in detail.

12.2.1 Cloud/edge-specific costs

The costs of operating applications purely on the cloud are expensive when compared to adopting edge-enabled sources. Notably, clouds have service-enabled offers to provide compute nodes, storage units, and networking devices. The service-based cloud developments and hosting features incur costs throughout the execution of applications. For instance, charges are calculated based on requests to deliver server instances as services or the amount of GB downloaded per application.

In fact, cloud-based services can lead to unnecessary costs apart from the utilization of resources for applications. There have been instances where IaaS services are not stopped properly. In some cases, the autoscaling feature of clouds increased the availability of resources for services, although the amount of resources was not required at times. Similarly, the services that involve Matlab or similar licensed software become complicated in assessing the cost utilization of services.

Additionally, including security-related products or allied software components to cloud-specific implementations can increase the costs associated with cloud-based offerings. In IoT-cloud applications, the desire to prevent security hazards is higher in tiny devices than at cloud nodes, as the sensor devices are more prone to security attacks. This is due to the fact that the tiny devices have memory or computational resource constraints.

To overcome a pure cloud-based sensor implementation, a serverless function-based execution model is adopted in modern edge-AI applications to improve costs to some extent. Especially, any IoT-based application does not require cloud services or associated resources throughout the period of the application. Instead, it needs to mark a few updates or avail decisions or suggestions depending on the data. To do so, a serverless-enabled execution model in the cloud will be a suitable, cost-efficient solution. This execution model can be deployed in transportation, healthcare, industries, or similar application domains by modifying the existing implementations.

The serverless execution model in the cloud involves data triggers to enable cloud hardware or associated services. It involves functions that could be configured such that the requirements of memory, compute, and network instances are specified. The execution model highly depends on the infrastructure providers, such as Amazon or Google. The entire implementation or infrastructure is triggered based on the functions and the business logics associated with the functions.

Serverless function-based application development requires a platform that considers the availability of resources. For instance, the function seeks to compute resources that might not be available on the platform at a particular time. In such situations, the server instances have to cold-start instances, i.e., the platform has to identify the right server where virtual instances could be created and booted for executing functions. The authors of [17] studied the impact of cold-start and warm-start instances while employing serverless platforms for applications.

Serverless platforms, being the execution environment of cloud providers, are prone to challenges that are posed by the providers. For instance, the auditing capabilities are hidden from users; also, selecting resources is bestowed on third-party providers on serverless platforms Hence, ample challenges exist in serverless-based implementations, although there are potential cost savings.

Edge-assisted applications have minimized the potential cloud-specific costs, including serverless-based cloud implementations. This is because distributed intelligence is offered using the amalgamation of several edge clusters with high resiliency. The security-related components could be offered by some reasonably capable in-house computing devices, such as Raspberry Pi nodes. These computing-capable nodes could be operated throughout the year without the intervention of clouds. Obviously, the costs associated with edge-enabled AI applications could be dramatically reduced.

12.2.2 Memory-related cost improvements

The cost associated with the memory usage of edge-AI applications can be fine-tuned depending on several factors. The management of memories and control features can increase the accessibility of data. Accordingly, the performance of applications increases. Hence, memory management and allied techniques can minimize the costs involved in edge-AI applications.

12.2.2.1 Caching

Usually, there is an approach to forwarding all requests and processing queries to clouds. This situation could be avoided in edge-enabled applications. The required models could be cached in edge clusters periodically. Also, the cloud-supported computing instances could utilize caching services. The authors of [9] revealed the importance of caches while integrating caching techniques into the edge-cloud platform. They studied the storage costs, remote data assessment costs, and other related costs of a video processing application.

The notion of including the caching component in edge-AI systems is to proactively cache cloud-computed learning models. Similarly, it could cache the most frequently utilized data for training or pre-training models in the working environment. A few cloud-specific caches that are developed and utilized in AI applications are AWS ElasticCache, Vertex-AI, Google CDN, Firestore, etc.

12.2.2.2 Memory management

Memory management is one of the primare tasks in system-level computing architectures. There are several memory management techniques in traditional computing architectures. They are memory allocation strategies, memory leakage prevention, memory garbage collection, proper memory references and trackers, memory deallocation, etc. Some techniques require the guidance of programmers or users, whereas others do not. The most explicitly discussed points in these memory management techniques are discussed below:

1. Memory allocation strategies: Memory allocation is often carried out by the operating system or user-defined conditions. Additionally, in recent years, software-defined memory allocation strategies have been widely studied by researchers. It is a better choice to effectively utilize memory units that are not located within the mother board. The costs could be dramatically reduced if memory utilization is elegantly handled on edge-enabled devices.

   Accessing memory units that are located away from an embedded device or board is a challenging task. There is the possibility of experiencing low speed in data transmissions. However, considering applications that do not require immediate processing, especially in IoT-enabled applications of various domains, the offboard memory units could be utilized to minimize the costs involved in operating applications.

   In fact, a common operating system such as FreeRTOS is not inherently designed to consider offboard memory units attached to edge-enabled devices. Hence, application developers are urged to extend the applications with sufficient instructions that consider the attached external memory units. This concept of developing edge-AI applications that can effectively utilize memory units that are not inbuilt and mounted on embedded devices is termed a software-defined memory allocation strategy. In recent years, several industrial applications or end-user applications have been developed based on the software-defined memory allocation strategy.
2. Memory leakage prevention: Memory leakage is a serious issue on any computing platform. It can be even worse in edge-enabled iterative applications because of increasing memory failures over a period of time. To prevent such memory leakage issues, either carefully allocate dynamic memories or avoid dynamism. The programmer has to look into potential memory leakage points in programs to avoid leakages. For instance, if malloc is utilized in programs, the programmer has to trace the path throughout the program execution prior to executing it. Similarly, it is better to utilize a dedicated performance analysis tool to assess memory health conditions if the dynamic memory will be allocated at runtime in an application.
3. Memory garbage collection: Similar to memory leakage identification, memory garbage collection is a memory management technique that automatically frees memory after the closure of programs. The garbage collection increases the costs involved in software practices. For instance, there will be high overheads in keeping track of memory outages and replacements when the size of the memory increases. Also, battery-operated devices will encounter communication failures due to the non-availability of memory sources.
4. Memory references and trackers: The modern features of over-the-air updates or similar kinds of advances pose potential risks to embedded devices. There are higher chances of wrong references to memory locations if applications are wrongly written or if they are challenged by security issues.

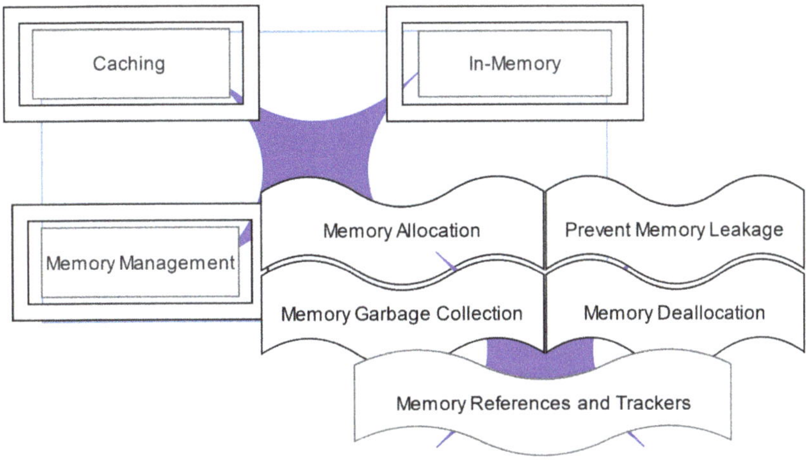

Figure 12.3. Memory-specific challenges in edge-AI systems.

5. Memory deallocation strategies: The memory allocation strategies differ in RTOS-based embedded devices when compared to server nodes. Similarly, the deallocation strategies are reverse operations. The deallocation processes have to be carried out immediately after the identification of garbage. Deallocation strategies in the majority of embedded devices are easier owing to static memory allocation processes. However, in modern applications, due to the modernization of processors and devices, the allocation and deallocation strategies of memories have to be handled in an effective manner.

Figure 12.3 provides a pictorial representation of memory-specific issues that hamper the business possibilities in edge-AI systems.

12.2.2.3 In-memory or high-end memory techniques

The in-memory technique is a technique that utilizes chip-level memory components or random access memory units to perform computations for edge-AI applications. The purpose of including the in-memory technique is to provide a faster computation step that dramatically decreases the operational costs of applications. For instance, cloud-based servers involved in edge-AI applications can service requests from edge nodes relatively better than traditional memory management techniques. Edge devices such as the Jetson Nano have 4–8 GB of RAM to compute a large volume of data-related operations in a short time. In addition to in-memory techniques, there are a few other memory techniques or hardware advances that can improve the operational costs involved in edge-AI applications.

12.3 Business opportunities: edge-AI platforms

Edge-AI platforms lead to increased business opportunities in traditional applications. The business increases in three broad dimensions: (i) software advances, and

(ii) hardware advances. This section explores the cost-improvement factors handled in various sectors due to the inclusion of edge-AI techniques.

12.3.1 Edge AI software: business opportunities

Researchers or practitioners develop software solutions that support edge-AI devices. In some cases, it is required to extend the existing cloud-only solutions to edge-cloud computing compendiums. This development, or conversion, has led to a business in several service-based industries. They create businesses on a large scale by supporting the businesses of others through the inclusion of edge-AI solutions. For instance, the traditional cloud-only solutions of predictive maintenance, text-conversion, human emotion detections, etc, are converted to edge-assisted AI solutions.

The advancement in software developments enables local data handling procedures with minimal costs, which can improve the costs involved in applications. Enterprises and other local business operators could minimize the utility of clouds by increasing the computations on edge-enabled devices. Similarly, offline computations using edge devices can avoid the need for networks throughout the year. Hence, many small business operations can be efficiently improved in terms of cost efficiency while applying edge-based solutions or converting the existing cloud-only solutions to heterogeneous solutions.

Similarly, software can be tailored to the needs of business operations. For instance, garment retailers can judge the preferences of customers immediately when they enter the retail shops. This could be judged by capturing the sensor data from wearables or garments. Also, predictions are carried out by learning the history of similar kinds of customers in the past. In the recent past, the application of ChatGPT or generative AI techniques has had multifold impacts on business operations.

By developing software for edge-AI systems, engineers or contributors can get associated with hardware developers or industries. This approach increases the manpower costs for edge-AI software developers. In fact, the capabilities of edge-AI engineers have to be increased by assessing the data and compute instances.

12.3.2 Edge AI hardware: business opportunities

In several solutions, edge-AI systems have been incorporated with sophisticated modern hardware assistance. For instance, video processing-related applications have been integrated with Jetson Nano or similar accelerator sticks for processing input data. By increasing the speed of processing inferences, users gain unique experiences that increase the business operations. However, the cost efficiency of the entire application will be improved.

Edge-AI hardware can increase the speed involved in business operations. As most of the operations in modern AI applications have been moved toward edge nodes, the costs involved in business operations have dropped, considering scalability, security, energy, and performance-related businesses.

The hardware advances enable digital twin operations or machine-to-machine communications to be easier when compared to the traditional approach of transporting data to cloud-centric computing units. A few specialized machines have been designed to establish buildings and designs in a large-scale in a manufacturing sector.

Similarly, the advancement of hardware manufacturing in edge-AI systems can minimize the compactness or utility of products. For instance, developing hardware units that are compact in size can increase the utility of computer vision-related products such as Google Glasses. The weight of the hardware units and their compactness can indirectly impact the businesses involved in their production.

12.4 Cost models

In this section, the costs involved in edge-AI applications are modeled depending on cloud resources and the associated execution models. Initially, it covers the costs of business operations if edge-AI applications are hosted using clouds. Finally, it reveals the importance of serverless functions to minimize the associated deployment costs.

12.4.1 Cloud-based cost models

Figure 12.4 lists some cost models that are derived when compute resources are mostly utilized in clouds. In clouds, it is expected to hire compute resources or servers using various pricing patterns such as pay-as-you-go, advertisements, subscriptions, and contributions.

The details of these types of cost models are described below:

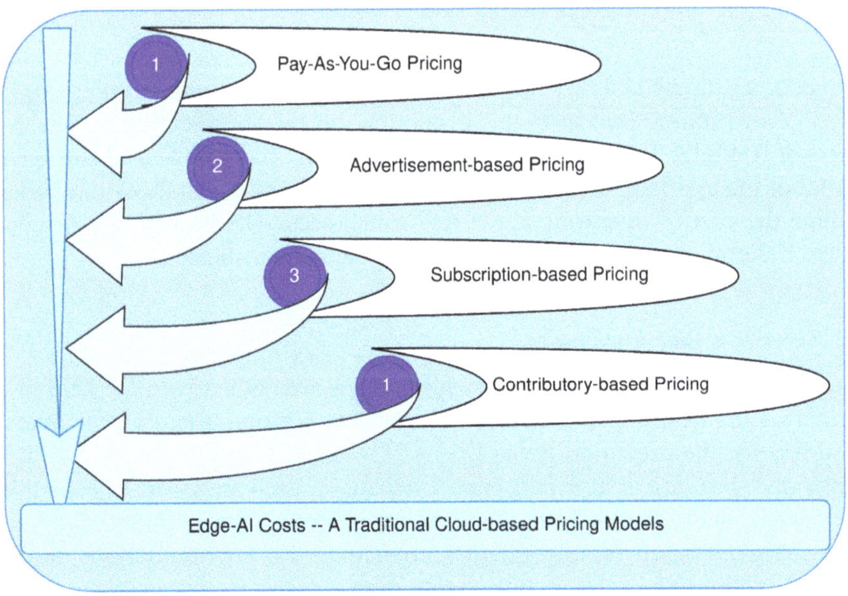

Figure 12.4. Cost models considering traditional clouds in edge-AI systems.

1. Pay-as-you-go pricing model: In this approach, the costs involved in executing edge-AI applications are calculated based on the edge computing resources that are utilized in the applications. This approach is similar to the cloud-based pricing strategy, in which all resources are assessed based on utilizations for specific applications.
2. Advertisement-based pricing model: In this approach, the charges are not levied on users of edge-AI applications. Instead, a few advertisement services are integrated into the application or services. The promotional approach can offer resources based on the computations that are compensated for by displaying advertisements.
3. Subscription-based pricing model: The subscription-based pricing model is a recurring pricing component that is charged for utilizing services or licensed products. Usually, the subscription-based pricing model tries to attract as many customers as possible by providing a lower subscription charge at the entry level. However, the resources are substantially charged over time based on the increasing demands.
4. Contributory pricing model: The contributory pricing model does not charge a fee for utilizing edge-level resources or any cloud-assisted resources in applications. Rather than charging a fee for resources, this pricing model allows the utilization of resources offered by contributory service providers. For instance, the resources belonging to edge nodes are utilized by cloud services.

The costs derived for Elastic Compute Cloud (EC2) of Amazon Web Services (AWS) is shown below:

$$\text{Cost}_{\text{EC2}} = \sum_{i} \tau_{\text{i}} \cdot c_{\text{i}}, \tag{12.1}$$

where i denotes the ith instance, τ_{i} the duration of the ith instance running, and c_i the hourly rate of running that instance, depending on the specification.

Users of AWS EC2 are typically charged according to runtime and storage units. The total of the expenses for both storing data and running instances can be used to determine the cost of operating these AWS instances. The hourly rate of compute instances determines the costs associated with running instances.

12.4.2 Serverless functions: costs

Function-as-a-service has emerged as a fascinating area of study in the field of cloud computing. Financially speaking, FaaS can be very alluring because consumers only have to pay for the execution time [4].

Generally speaking, FaaS has been extensively used in several fields utilizing public clouds, such as cloud robotics. As an example, the authors of [3] and [6] created sampling-based motion planner algorithms for robotic actions employing serverless functions and robotic motion planning algorithms, respectively. Similar to this, the authors of [14] created data pipeline functions to carry out machine learning algorithms.

Because the deployment strategy is widely used across several sectors, researchers looked into how various applications affected performance. Notably, AWS Lambda can be 600 times less expensive for very low consumption than AWS EC2 t2.nano, according to the authors of [1]; yet, it can also cost significantly more than Google App Engines.

The authors of [16] investigated three different system designs for a web application: (i) a microservice architecture, (ii) a monolithic architecture, and (iii) FaaS. Their test results show that using FaaS can reduce infrastructure costs by over 50% compared to microservice architecture and over 70% compared to monolithic design. Similar to this, the creators of [15] used distributed streaming functions and FaaS to execute edge-related stateless or stateful activities. The work investigated the impact of related expenses on event-based processing. However, the article does not include an analytical perspective for evaluating the costs related to serverless operations by public cloud providers.

An economic model for FaaS that can roughly estimate the overall cost of deploying AWS Lambda on EC2. To this end, models are created and loads are simulated to evaluate the costs of functions. It was close to the point at which EC2 and AWS Lambda break even. The authors of [11] devised a number of cost-optimization algorithms for FaaS-enabled applications. The authors of [7] learned about the features of FaaS-based media applications driven by 5G.

For several cloud deployments, Tayal *et al* from Deloitte Consulting [5] took advantage of the cheaper infrastructure costs in addition to the development and maintenance costs. The total cost of ownership (TCO) is the sum of those three expenses. FaaS can have a lower total cost of ownership (TCO) than standard IaaS because of its reduced development and maintenance expenses, which offset any higher infrastructure costs.

The performance of FaaS has been the subject of several examinations by application developers. Request latency was assessed on AWS Lambda and found to be higher than on AWS Elastic Beanstalk (PaaS) by [5]. AWS, Azure, and Google's FaaS solutions were compared by McGrath and Brenner [13]. They discovered that, compared to the other two methods, the performance of the AWS solution was superior. AWS Lambda has more difficulty serving `GET` and `POST` requests as API endpoints to query a database.

In 2020, Mahmoudi *et al* introduced an analytical performance model [12]. Their method is able to forecast a number of serverless computing performance measures, including average latency, average running/idle instances, average usage, etc. The inherent workload characteristics of serverless functions were overlooked by this technique. Furthermore, the writers of [8] have just created cost-optimization plans for AWS serverless functions. Their strategy centered on using heuristic techniques to reduce the costs associated with AWS's serverless capabilities.

One of the most important issues in edge-AI systems created with functions is modeling the costs associated with serverless functions. This section details the associated expenses for various services, with a particular focus on serverless platforms offered by Amazon AWS Lambda. There are various ways to handle incoming requests in order to serve them on the server. For instance, we can use AWS Lambda function instances, or we can use several EC2 instances. The cost is associated with each of the deployment techniques. The deployment strategy and its theoretical cost are denoted by the phrase *cost model.*

The cost of using AWS Lambda is formulated as follows:

$$\text{Cost}_{\lambda} = \sum_{i} m_{\text{i}} \cdot d_{\text{i}} \cdot c_{\lambda} + r, \tag{12.2}$$

where i denotes the ith function instance, m_i the memory allocated to that instance in GB, d_i the duration of that instance running in seconds, c_{λ} the cost rate of AWS Lambda, and r the fixed cost per request.

AWS Lambda is a FaaS offering from AWS. Users of AWS Lambda are charged according to the quantity of memory utilized in gigabytes (GB) in a second and the number of requests for function invocations. Both the code's initialization and execution are included in the duration. In addition to allocating memory to a function instance between 128 MB and 10,240 MB in 1 MB increments, the user can select between an x86 or an ARM architecture and measure the time with a resolution of 1 ms. Usually, AWS Lambda charges $0.000 016 6667 for every GB-second on the x86 architecture, $0.000 013 3334 on the ARM architecture, and $0.20 for every million requests.

The edge-AI that utilizes AWS Lambda services will include communication delays while calculating the costs involved in the operations. The costs are derived as follows:

$$\text{Cost}_{\lambda-e} = \sum_{i} m_{\text{i}} \cdot d_{\text{i}} \cdot c_{\lambda} + r + C, \tag{12.3}$$

where C denotes the communications involved in processing function-oriented tasks.

12.5 Economic simulators for FaaS implementation

Although there are a few stand-alone studies, there aren't many that advise edge-AI users on the economics of serverless functions in public clouds like AWS Lambda while taking function-related factors like memory requirements into account.

One of the most important things for cloud application developers to consider is the expense of implementing serverless functionalities. This section details the functional components involved in the development of an Economic Cost Evaluation Framework (ECEF) for edge-AI systems. The ECEF framework contains a request-making component, queue, function instances, and nonitoring servers, for calculating the potential costs of edge-AI applications (see figure 12.5).

The details of these components are described in the following subsections.

12.5.1 Request-making component

The ECEF's `request-making component` creates requests or function calls. Users can select requests using probability distributions or at random. To select

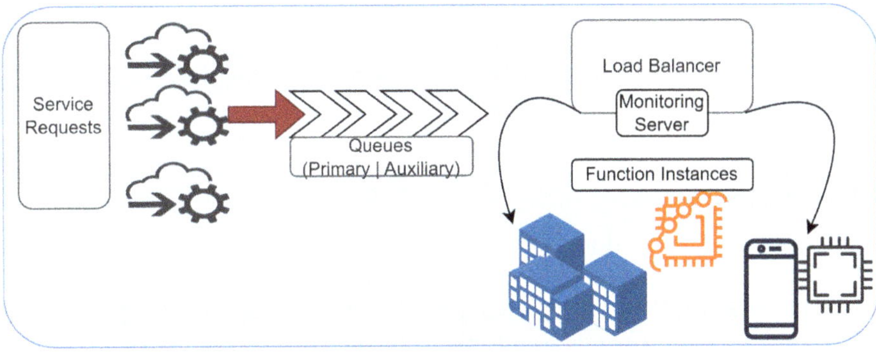

Figure 12.5. ECEF framework for evaluating costs in edge-AI.

several requests to be issued in a period, for example, users of ECEF could select the Poisson distribution with parameter λ. Additionally, the RAM needed for each of those requests could be specified by the users. Moreover, users have the option to combine many of these generators into a single request, with each request containing λ parameters and a specific memory requirement. `Queues` is the name of the requests generated by the `request-making component`.

12.5.2 Queues

FaaS-based edge-AI applications may make use of a number of autoscaling techniques, including *scale-per-request*, *concurrency value scaling*, and *metrics-based scaling*. In the *scale-per-request* approach, a request can be processed by initializing a new function instance with sufficient memory. In the event that no additional function instances are available, these requests are handled. A function instance handles several requests at once in *concurrency value scaling*. Finally, there is *metrics-based scaling*, which is a rule-based approach that maintains certain metrics, such as CPU or memory consumption, within a predetermined range. The *scale-per-request* pattern is employed by the majority of well-known FaaS vendors.

The working of ECEF is predicated on the *scale-per-request* approach. A FaaS-based system can have an unlimited number of function instances. Of course, it might launch a fresh instance at any time. However, a server is used to limit the amount of instance invocations that can occur at once in order to control extreme situations.

The two halves of the queue that run concurrently are the primary and auxiliary queues. One of these is the initial queue for all requests coming from the request-making component. The load balancer can choose from among the function instances that were initially created and made available in order to respond to that kind of request. If it finds one instance of the function, it may assign the request to it. If it finds none and the function instance limit is not yet reached, it will initialize one with enough memory for the request. Requests are moved to the other queue if they are not.

Next up for the load balancer is the next request. If the current queue is empty, the load balancer switches to the other queue and transmits incoming requests. These two queues are designed to make sure that when the load balancer allocates a function instance, it gives precedence to the waiting requests.

12.5.3 Function instances

The real computation is done by the `function instance` component of the ECEF. This part is utilized to function as intended within AWS Lambda. The common states `STARTING`, `RUNNING`, and `READY` are used to accomplish it.

A `function instance` goes into the *cold-start* phase when it is first initialized. A `function instance` can only go through this phase once in its lifespan, and it can take the form of the `STARTING` state. The `function instance` is unable to process any load balancer requests at this time. Moreover, the `function instance` costs money and uses memory.

This phase includes both the process where the infrastructure starts up a new virtual machine or container and *application initializing*. The *application initializing* process encompasses initial tasks performed by the user's code, such as creating a database connection, downloading reference data, or loading heavy frameworks such as Spring Boot. Although only the application initializing is billed, both steps are billed in this simulation. This phase can last anywhere from a couple of milliseconds to seconds, depending on the language runtime and memory size. However, these effects are considered uniform in the simulation.

It is possible to modify the simulation's `function instance` memory usage by adjusting its parameters in relation to the memory that is available. Typically, that area of memory has a distribution of memory consumption. For instance, the cold-start consumes 40% of the 256 MB of RAM allocated to the `function instance`. After then, the `function instance`'s memory usage during the cold-start is similar to a normal distribution with a center at $\lfloor 0.40 \times 256\text{MB} \rfloor = 102\text{MB}$. A certain amount of standard deviation is acknowledged and is modifiable within the FES parameters.

When there is an unfulfilled request and the function instance limit has not been reached, the ECEF method initializes a `function instance`. The function instance's state changes to `RUNNING` and the request is instantly assigned to it upon the completion of the initialization process. It is also possible to modify the memory used in this mode using the options. It is also possible to apply linear regression to an exponential model to approximate the execution time from empirical testing, since the runtime is dependent on the memory available on the instance.

Every function that originates from the request-making component is assumed to be unbounded by IO operations. The CPU is its boundary. Upon completion of processing a request, the `function instance` goes into a cool-down phase, or the `READY` state. The load balancer can revert to the `RUNNING` state during this phase and can also assign a new request to the `function instance`. The `function instance` is terminated if it remains in this condition for an extended amount of time without receiving a new request. The time spent in this state is not

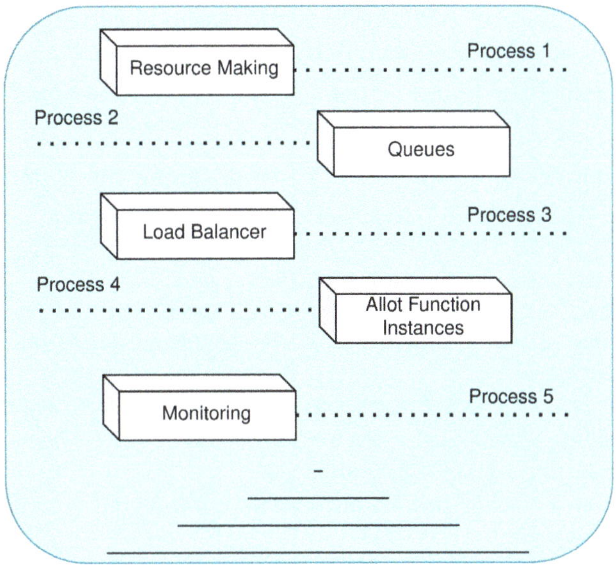

Figure 12.6. Processes and software components for simulating costs of edge-AI.

billed, in contrast to the other two states. A `function instance` may idle for up to an hour before being shut down; this behavior is non-deterministic.

12.5.4 Monitoring server

The `monitoring server` of ECEF does the following activities:

1. it holds a pool of function instances that process the incoming requests from the queue;
2. it gathers a few performance metrics; and
3. it calculates several cost models.

When a function instance completes processing a request, the starting and ending timestamps, running and waiting times, and the function instance itself may all be logged by the monitoring server. The component may have a limit on the number of instances that run concurrently. Figure 12.6 represents the entire process involved in simulating the costs of edge-AI systems and the associated software components in a pictorial form.

12.6 Conclusion

There are ample ways of improving the cost involved in operating edge-AI applications. For instance, avoiding clouds or engaging edge-based devices can minimize the costs of applications. However, there is a need to list the potential business opportunities due to the involvement of edge-AI systems. This chapter showcases a few potential use cases to improve the business opportunities and cost factors of edge-AI applications. It delves into deriving the economic cost model of applications while utilizing AWS-based cloud setups. Learning the different cost models can enable edge-AI users to rethink utilizing resources, considering the costs involved in applications.

References

[1] Adzic G and Chatley R 2017 Serverless computing: economic and architectural impact *Proceedings of the 2017 11th Joint Meeting on Foundations of Software Engineering* (New York, NY: Association for Computing Machinery) pp 884–9

[2] Akter S, Michael K, Uddin M R, McCarthy G and Rahman M 2022 Transforming business using digital innovations: the application of AI, blockchain, cloud and data analytics *Ann. Oper. Res.* **308** 7–39

[3] Anand R, Ichnowski J, Wu C, Hellerstein J M, Gonzalez J E and Goldberg K 2021 Serverless multi-query motion planning for fog robotics *2021 IEEE International Conference on Robotics and Automation (ICRA)* pp 7457–63

[4] Eivy A and Weinman J 2017 Be wary of the economics of 'serverless' cloud computing *IEEE Cloud Comput.* **4** 6–12

[5] Arora N J G, Choi W and Sahoo B K 2023 Determining the total cost of ownership: comparing serverless and server-based technologies (https://d1.awsstatic.com/psc-digital/2023/gc-300/deloitte-tco-mod/determining-the-total-cost-of-ownership.pdf) (accessed 31 January 2024)

[6] Ichnowski J, Lee W, Murta V, Paradis S, Alterovitz R, Gonzalez J E, Stoica I and Goldberg K 2020 Fog robotics algorithms for distributed motion planning using lambda serverless computing *2020 IEEE International Conference on Robotics and Automation (ICRA)* pp 4232–8

[7] Konstantoudakis K, Breitgand D, Doumanoglou A, Zioulis N, Weit A, Christaki K, Drakoulis P, Christakis E, Zarpalas D and Daras P 2022 Serverless streaming for emerging media: towards 5G network-driven cost optimization *Multimedia Tools Appl.* **81** 12211–50

[8] Kumari A, Sahoo B and Behera R K 2023 Workflow aware analytical model to predict performance and cost of serverless execution *Concurrency Comput. Pract. Exp.* **35** e7743

[9] Li C, Song M, Yu C and Luo Y 2021 Mobility and marginal gain based content caching and placement for cooperative edge-cloud computing *Inf. Sci.* **548** 153–76

[10] Li C, Sun H, Chen Y and Luo Y 2019 Edge cloud resource expansion and shrinkage based on workload for minimizing the cost *Future Gener. Comput. Syst.* **101** 327–40

[11] Lin C, Mahmoudi N, Fan C and Khazaei H 2023 Fine-grained performance and cost modeling and optimization for FaaS applications *IEEE Trans. Parallel Distrib. Syst.* **34** 180–94

[12] Mahmoudi N and Khazaei H 2022 Performance modeling of serverless computing platforms *IEEE Trans. Cloud Comput.* **10** 2834–47

[13] McGrath G and Brenner P R 2017 Serverless computing: design, implementation, and performance *2017 IEEE 37th International Conference on Distributed Computing Systems Workshops (ICDCSW)* pp 405–10

[14] Paraskevoulakou E and Kyriazis D 2023 ML-FaaS: toward exploiting the serverless paradigm to facilitate machine learning functions as a service *IEEE Trans. Netw. Serv. Manag.* **20** 2110–23

[15] Pfandzelter T, Henning S, Schirmer T, Hasselbring W and Bermbach D 2022 Streaming vs. functions: a cost perspective on cloud event processing *Proc. of the 10th IEEE Int. Conf. on Cloud Engineering (IC2E 2022) (Pacific Grove, CA)* pp 67–78

[16] Villamizar M *et al* 2016 Infrastructure cost comparison of running web applications in the cloud using AWS lambda and monolithic and microservice architectures *2016 16th IEEE/ACM International Symposium on Cluster, Cloud and Grid Computing (CCGrid)* pp 179–82
[17] Xiao K, Yang S, Li F, Zhu L, Chen X and Fu X 2024 Making serverless not so cold in edge clouds: a cost-effective online approach *IEEE Trans. Mobile Comput.* pp 1–14

IOP Publishing

Edge Intelligence
Deep learning-enabled edge computing
Shajulin Benedict

Chapter 13

Challenges and future directions

This chapter describes the potential challenges and research directions that could be addressed in the edge-AI research domain. It summarizes previous chapters, with more emphasis on identifying research challenges. The chapter provides a holistic picture of future implementation directions so that researchers or readers can start to develop their works by solving the identified problems in their respective domains.

13.1 Edge-AI challenges

Applying intelligence using edge-enabled devices and scalable computing devices, including clouds, has increasing challenges. A number of challenges have been identified by researchers and practitioners. Accordingly, they are targeted by researchers as part of their PhD program, academic activity, or industrial assignments.

We have learned about several topics in this book that explicitly cover edge architectures, edge-enabled learning methodologies, edge-based inference techniques, device-level accelerators, performance concerns, security features, edge-AI frameworks, orchestration tools, edge-assisted applications, and business opportunities.

Now, we will understand the potential challenges identified in these topics so that researchers or future developers can directly target these research problems.

13.1.1 Energy consumption issues

Edge-enabled devices are attached to battery-operated sensors and actuators. Often, these devices are computationally inexpensive and have to be handled with a limited energy source. Or, these devices have to be powered by renewable energy sources. Obviously, the requirement to reduce the energy consumption of these compute units has dramatically increased among the minds of researchers and practitioners.

In chapter 7, the importance of handling energy consumption issues in edge-AI applications has been discussed in detail. Additionally, some solutions that have been practiced in the past have been studied.

doi:10.1088/978-0-7503-5593-3ch13

13.1.2 Resource constraints

Edge devices are connected with tiny sensor units. Often, these devices are designed to support portability, i.e., the devices are lightweight and size-sensitive. Similarly, the devices are hosted to support multiple applications from different administrative domains. For instance, Jetson Nano boards are utilized by weather prediction algorithms or federated learning algorithms. This means that multiple virtual environments can be experimented with on these nodes. Hence, resource constraints are considered a challenging aspect of performing edge intelligence.

The most seriously discussed resource constraints in the edge intelligence domain are explained in the following subsections.

13.1.2.1 Memory constraints

It is a known fact that edge-enabled devices are not powerful enough owing to space limitations. The hardware is designed to accommodate a limited number of memory units so that it can be lightweight. For instance, modern Jetson Nano devices have support for 4 GB of RAM compared to cloud servers which have unlimited memory space. Obviously, the number of applications that could be executed on these devices will be a challenge. The associated operating systems have to carefully launch programs into the restricted memory footprints. Researchers have been working on the memory-related issues of edge-AI applications so that several edge-AI applications could be managed with a limited memory unit.

13.1.2.2 CPU/GPU/TPU constraints

Designing powerful CPU/GPU/TPU computing components for edge-AI applications is not a viable approach. This is because of the limited space and energy concerns of edge devices. The edge devices have to accommodate several other components in addition to computing components, especially the I/O handling units. Although technologies have improved in recent years to enhance the computing capability of computing units, they are prone to heat dissipation issues and energy consumption issues.

Therefore, industrialists or researchers have suggested several process management techniques associated with operating systems. For instance, Intel has recently developed a fanless computing node. Notably, Intel's products, such as the E50 Palm-size IoT Gateway/Servers, the E100 Ultra-compact Fanless Servers, and the E302 Mini-ITX Compact Fanless Servers, have attracted energy-conscious edge-AI designers as fans consume a large portion of battery usage or electric power.

Due to the non-availability of certain computing power, edge-AI applications are prone to poor learning accuracy. At times, the algorithms have to be optimized, considering the availability of resources.

13.1.2.3 Network constraints

In general, clouds have sufficient network bandwidth to execute distributed applications. For instance, data centers have multiple rack servers that are connected to each other using infiniband network switches instead of the traditional gigabit switches.

However, edge devices, while forming edge-level clusters, cannot increase the network bandwidth. This limitation has impacted edge-AI applications. The applications, therefore, have to submit the data to powerful compute clusters that are geographically away from edge nodes. Consequently, the applications encounter higher latency and privacy concerns.

To override these challenges, a number of solutions have been proposed in the past. However, the solutions could be enriched with more novel device management techniques or resource optimization solutions. For instance, software-defined networking components could be introduced at the edge level to increase the network bandwidth of entire edge-AI applications.

It is also crucial to handle the interoperability issues evolving in the networking devices of tiny components. Notably, developing automated networking based on software defined networking is possible in clouds. However, it can lead to resource constraint-related issues in edge-enabled AI systems.

13.1.3 Performance efficiency

Although a few performance improvement suggestions and solutions have been designed in the past, there are several opportunities that could be adopted to improve the performance of edge-AI applications.

13.1.3.1 Improved learning

Involving edge clusters or subsets of computing nodes to learn environments with higher accuracy is an optimization problem. For instance, increasing the number of compute nodes could increase the learning accuracy due to the engagement of a large volume of data. However, considering resource utilization and other factors, the end outcome of the edge-AI application could be in vain.

Researchers have adopted numerous technologies to adopt learning approaches involving edge devices. For instance, federated learning, adaptive federated learning, transfer learning, gossip learning, collaborative modeling, cooperative modeling, etc. A detailed explanation of these learning and inference techniques has been discussed in chapters 4 and 5.

In fact, improving the data processing methods, pre-processing methods, data handling methods, data accessing procedures, etc, invites technical challenges. There are possibilities for pursuing ample research in these areas. Notably, there is a huge possibility for inconsistency of data if security is hampered in learning processes.

Additionally, researchers could work on developing novel learning algorithms that improve learning accuracy. They can design a few architectural changes while executing learning algorithms. For instance, a novel transfer learning platform or federated learning platform is one approach to improving the learning processes of edge-AI applications. In addition, novel quantization procedures or compression techniques could be adopted considering the communication hurdles among edge nodes; the anomalies or outliers of learning modes could be tracked in time.

In some transfer learning or federated learning environments, the edge devices are exposed to dangerous working environments. For instance, edge devices that are housed in industrial plants, including weather monitoring stations, are openly mounted on machines that are exposed to hot sunlight. Obviously, there is a possibility of temperature variations in the underlying electronic devices that hamper the learning accuracy of edge-AI applications. It could even reduce the life span of computing nodes, which leads to varying networking topologies and reduced computing nodes while learning and reasoning about the environments.

In the future, learning algorithms could be designed considering the explainability nature of algorithms. The explainability level could be increased by adopting robust techniques and implementation packages. In fact, researchers have been developing a few explainability-related learning algorithms. However, they are not sufficient or adapted to all working edge-AI environments.

13.1.3.2 Decision making

Decisions in selecting futuristic options and learning environments could be improved by involving real-time actions. For instance, deploying models on resource-constrained edge nodes can be restricted to a subset of compute nodes based on the data affinity and privacy concerns of the environment. The decisions toward creating actions need to be made in near-real-time. This has to deal with situations such as communication latency between edge nodes and the decision time of algorithms.

Making decisions can be a challenging endeavor when the resources belong to different administrative domains. This involves enabling firewall settings that favor the underlying applications that are distributed across regions. Also, the decisions have to consider the heterogeneity involved in compute nodes. Notably, the architectures of Jetson Nano and Raspberry Pi nodes are different from ESP32s. Engaging different heterogeneous edge nodes in edge-AI applications challenges the frameworks or orchestration tools to make agile decisions, i.e., the decisions need to consider several parameters before making an appropriate suggestion.

There is a need to adopt a well-balanced computing continuum that surpasses and offloads computing from edge to the cloud environments. The supporting nodes and clusters will have to scale automatically depending on the load involved in the learning processes. Obviously, there exist a number of challenges in adopting a robust load balancing algorithm and in developing inter-cloud computing environments while executing any edge-AI applications.

In some situations, there is no necessity to increase the performance of edge-AI applications due to poor latencies or computational delays exhibited on specific edge nodes. Hence, the decisions could be delayed considering the underlying edge-based compute nodes and the scalable cloud servers for edge-AI applications.

13.1.4 Secured learning

Privacy and identity are two crucial security-related pillars in providing edge AI. The existing solutions that could offer data privacy and protect the authenticity of edge devices are not sufficient due to tiny devices. Also, edge-enabled devices are prone to physical attacks in addition to several security attacks that are widely discussed in chapter 8. Physical attacks are more commonly found while mounting sensor nodes in vulnerable locations.

The cost involved in edge-AI applications could increase depending on the utilization of security algorithms. In fact, there is a need for solutions that could automatically manage the security requirements of edge-AI applications.

Deploying models or transferring learning models using lightweight modeling algorithms is a challenge in several application domains, including healthcare-related AI solutions. Designing frameworks that handle offloading procedures without security-related issues on multiple edge devices is a seriously adopted research problem in the edge-AI domain. For instance, the models derived from a local dataset can pose security hazards to global models when the data are hampered with poor-quality information.

13.1.5 Legal and standards

Edge-AI faces several challenges and poses opportunities due to legal or standard failures. Notably, utilizing images or video files of people can attract large penalties if they are used without permission. There have been cases filed by people to prevent the actions of copying or leaking data. Unfortunately, edge-AI requires apt data to establish learning models and augment intelligence using edge or cloud server instances.

These legal advice differs geographically across countries. For instance, applying human images to developing technologies is not a serious concern in countries such as India and China. However, the situation could attract penalties or imprisonment in the USA and countries in Europe.

The legal policies finalized between countries or organizations need not be a long-term rule. Accordingly, edge-AI applications that have been running for years will have issues, especially when the applications need to be treated during the occurrence of errors or failures. Due to such failures, if frameworks adopt resource sharing policies or data sharing policies, the accuracy of the learning system will be a challenge. Sometimes, the organizations involved in the distributed computing processes can hold data or learning models, which leads to vendor lock-in issues.

The lack of standards while coordinating or integrating edge-AI or while sharing resources opens up new challenges. Researchers have been developing several standalone solutions that address these issues. If common standards have been developed for executing edge-AI applications, the dynamic resource allocation procedures will be handled with ease considering different edge-AI workloads; also, the real-time installation of resources will be faster than the ongoing approaches.

www.ingramcontent.com/pod-product-compliance
Ingram Content Group UK Ltd.
Pitfield, Milton Keynes, MK11 3LW, UK
UKHW051859100926
8556IPUK00005B/695

9 780750 355919